Investigating I to Resolve Environmental Challenges

Proceedings of the
Twenty-Eighth International Conference
on the Unity of the Sciences

Global Online Conference
April 13 –14, 2022

Hyo Jeong International Foundation
For the Unity of the Sciences Press

This Proceedings volume contains content delivered at the
Twenty-Eighth International Conference on the Unity of the Sciences
(ICUS XXVIII). This book is the Second Edition.

Copyright © 2023 by Hyo Jeong International Foundation
for the Unity of the Sciences (HJIFUS) Press

All rights reserved by HJIFUS Press, a shared project of HJIFUS, USA,
and HJIFUS-Korea. Except for use in reviews, direct use by the authors,
or use with the prior written permission of the publisher, no part
of this book may be reproduced in any written, electronic, recorded,
or photocopied form or via other media of expression now known
or later developed.

The views expressed herein are those of the authors and do not
necessarily represent the views of HJIFUS, USA, or HJIFUS-Korea,
or any of their founders, directors, officers, or staff, or any
of their sponsors.

Book design by Gil Roschuni
Front cover photo by Jarosław Ponikowski from Pexels.com
Back cover photo by Avery Nielsen-Webb from Pexels.com
Section fronts photo by Marita Kavelashvili from Unsplash.com

Library of Congress Control Number: 2023909356
ISBN: 978-1-947499-14-0

Published by
Hyo Jeong International Foundation for the Unity of the Sciences Press

3600 New York Ave. NE	324-211 Misari-ro, Seorak-myeon
Washington DC 20002	Gapyeong-gun, Gyeonggi-do
USA	Republic of Korea

Table of Contents

    v | Preface

OPENING PLENARY SESSION

    3 | Welcoming Remarks: *Dr. Douglas D.M. Joo*
                   (Chair, ICUS XXVIII)

    6 | Congratulatory Address: *Dr. Jeong Kee Hong*
                   (Vice-Minister of the Environment, Republic of Korea)

    8 | Founder's Address: *Dr. Sun Jin Moon*
                   (Representing the Founders)

  13 | Keynote Address: *Prof. David MacMillan*
                   New Catalytic Strategies for a Sustainable Future

SESSION 1

Addressing Climate Change: Strategies to Achieve "Net Zero"

  30 | About Session 1
                   Chair: *Prof. Cliff Davidson*

  31 | Presentation One: *Prof. David Blekhman*
                   Tapping Hydrogen for Electric Power and Transportation

  50 | First Commentary: *Prof. Kazunari Domen*

  55 | Second Commentary: *Dr. Dave Edlund*

  60 | Presentation Two: *Dr. Eric Larson*
                   Negative Emissions Technologies in U.S. Decarbonization Pathways

  74 | First Commentary: *Prof. Larry Baxter*

  81 | Second Commentary: *Prof. Steven Chuang*

  84 | Discussion

SESSION 2

Manufacturing Materials for Eco-Friendly Products

  92 | About Session 2
                   Chair: *Prof. Michael Stenstrom*

  93 | Presentation One: *Hon. Mike Lancaster*
                   The Promise of Green Chemistry

 111 | First Commentary: *Prof. James Clark*

 115 | Second Commentary: *Prof. Bimlesh Lochab*

 124 | Presentation Two: *Prof. Michael Shaver*
                   Rethinking the Plastics Revolution

 132 | First Commentary: *Prof. Mark Miodownik*

 135 | Second Commentary: *Dr. Carly Fletcher*

 139 | Discussion

SESSION 3
Engaging the Public in Tackling Environmental Concerns

- 146 | About Session 3
    Chair: *Prof. Suh-Yong Chung*
- 147 | Presentation One: *Dr. Kazuo Matsushita*
    Policy Challenges to Achieve a Net-Zero Society by 2050: A Perspective from Japan
- 157 | First Commentary: *Prof. Wil Burns*
- 163 | Second Commentary: *Prof. Marilyn Brown*
- 165 | Presentation Two: *Prof. Bruce Johnson*
    Promoting Grassroots Action on Environmental Issues
- 177 | First Commentary: *Prof. Dilafruz Williams*
- 181 | Second Commentary: *Prof. Megan Bang*
- 186 | Discussion

CLOSING SESSION
- 196 | Reports by the Three Session Chairs: *Prof. Cliff Davidson, Prof. Michael Stenstrom,* and *Prof. Suh-Yong Chung*
- 205 | Closing Remarks: *Dr. Douglas D.M. Joo*

APPENDIX
- 208 | Organizing Committee and Staff
- 209 | List of Attendees
- 215 | A Brief History of ICUS
- 218 | Founders of ICUS and HJIFUS
- 222 | Introduction to HJIFUS
- 226 | HJIFUS Boards of Directors
- 227 | Sponsoring and Supporting Organizations and Worldwide HJIFUS Offices
- 228 | Index

Preface

I am very pleased to present the proceedings of the Twenty-Eighth International Conference on the Unity of the Sciences (ICUS XXVIII), held on April 13–14, 2022 (Korea time). The conference was sponsored by the Hyo Jeong International Foundation for the Unity of the Sciences (HJIFUS), organized as a joint project of HJIFUS offices in Korea and the United States, and supported by related offices in Japan and Europe.

The ICUS legacy stretches back to 1972, when the series was founded by Rev. Dr. Sun Myung Moon and Rev. Dr. Hak Ja Han Moon. The period from 1972 to 2000 marked the first phase of the series, which promoted unity among the sciences guided by absolute values—or values that benefit all of humanity—to formulate ways to resolve many critical issues. After 22 conferences, the series was suspended in 2000, when interdisciplinary collaborations were becoming more common.

By the year 2017, however, environmental troubles were growing at an alarming rate. In response, Rev. Dr. Hak Ja Han Moon launched the second phase of the ICUS series, placing the focus on identifying scientific and technological solutions to these problems. Thus, in 2017, ICUS XXIII was held with the theme, *Earth's Environmental Crisis and the Role of Science;* in 2018, ICUS XXIV had the theme, *Scientific Solutions to the Earth's Environmental Challenges;* in 2019, ICUS XXV had the theme, *Environmental Health and the Quality of Human Life;* in 2020, the theme for ICUS XXVI was, *Resolving Environmental Threats for the Benefit of Humanity;* and in 2021, the theme for ICUS XXVII was, *Surmounting the Challenges of Environmental Stressors.*

This year (2022), the theme for ICUS XXVIII was, *Investigating Pathways to Resolve Environmental Challenges.* There were 90 registered attendees representing nine countries. The conference consisted of an opening session, three academic sessions, and a closing session.

The opening session included a keynote address by Professor David MacMillan of Princeton University, who received the

2021 Nobel Prize in Chemistry. He outlined the development of novel catalyzed reactions that can be used to synthesize useful products at high efficiency, simultaneously enhancing environmental sustainability.

The first academic session described how "net zero" of carbon emissions can be achieved by using green hydrogen, or by employing biological processes or carbon capture technologies. In the next session, the emphasis was on the importance of "green chemistry," and the benefits and drawbacks of plastics. The final academic session showed how policy making and education can help engage the public in taking action. The text is generally illustrated with figures, as seen in the keynote address, and in the presentation on hydrogen.

It is our hope that the valuable information presented at ICUS XXVIII will be widely shared, making a genuine impact on the resolution of environmental challenges around the globe. We therefore plan to distribute these proceedings to educators, researchers, policy makers, libraries, university departments, environmental organizations, and others committed to protecting and restoring the health of our planet. We will also post the presentations on our website.

At the end of this book, the Appendix contains information about the conference organizing committee and staff, a list of attendees, and a brief history of ICUS. There is also some background information on the founders of ICUS and HJIFUS. In addition, there is a short introduction to HJIFUS, with the boards of directors and contact information for our worldwide offices.

In closing, I thank all those who helped make ICUS XXVIII an exceptional success, including the session chairs, main speakers, commentators, other participants, organizing committee members, and staff. I also thank those who invested their efforts to produce this book. ◊

Dr. Douglas D.M. Joo
Chair, ICUS XXVIII and HJIFUS

ICUS XXVIII

OPENING PLENARY SESSION

Investigating Pathways to Resolve Environmental Challenges

Dr. Douglas D.M. Joo, Chair of ICUS XXVIII, delivers welcoming remarks

Dr. Sun Jin Moon gives a special address on behalf of Rev. Dr. Hak Ja Han Moon, Cofounder of ICUS.

Prof. David MacMillan gives the keynote address.

ICUS XXVIII participants tune in for the opening session.

Opening Session
Welcoming Remarks
Dr. Douglas D.M. Joo
Chair, ICUS XXVIII

Respected Scientists, Scholars, Ladies and Gentlemen:

On behalf of the Hyo Jeong International Foundation for the Unity of the Sciences (HJIFUS), I am very pleased to welcome you to the Twenty-Eighth International Conference on the Unity of the Sciences (or ICUS XXVIII). This year, the conference theme is, *Investigating Pathways to Resolve Environmental Challenges.*

As you may know, HJIFUS is dedicated to finding, promoting, and implementing the most promising solutions to environmental problems. Toward this end, we have been sponsoring annual conferences in the ICUS series since 2017.

I should note that the ICUS series was founded in 1972 by Rev. Dr. Sun Myung Moon and Rev. Dr. Hak Ja Han Moon. At that time, the underlying motif was to promote unity among the sciences, guided by absolute values, that is, values that benefit all humanity. However, with the founding of HJIFUS in 2017, the focus of the ICUS meetings was directed toward addressing environmental issues, mainly through conventional scientific approaches. Perhaps now is a time to broaden our search by probing the frontiers of science that may involve unconventional ideas and approaches.

Last year, on account of travel restrictions imposed by the pandemic, we held ICUS XXVII as an online meeting. Once again, for this year's meeting, we are using a virtual platform to circumvent the pandemic-related restrictions in international travel. I wish to thank all of you, distinguished guests, for overcoming the difficulties imposed by the pandemic, including the need to transcend differences in time zones, to attend this conference.

The ongoing COVID-19 pandemic and environmental issues such as climate change have one thing in common: they are global in scope. They point to the fact that all people are interconnected with one another and with nature. Therefore, to tackle these problems, the people of the world need to work together, cooperatively. This is also a reason why these conferences need to be held on a global scale.

Our founders have called for the equalization of technology around the world. In this case, as new technological solutions to environmental problems are discovered and developed, they need to be shared with all people in all nations. In this manner, we need to follow the principles of interdependence, mutual prosperity, and universal values, which will lead to world peace.

Over the past six years, we have been very fortunate to have had excellent speakers covering a wide range of urgent topics. For example, Prof. Mario Molina has spoken about the seriousness of climate change, and Prof. Luc Montagnier has covered health issues, including the COVID-19 pandemic. In addition, we have had expert presentations on topics such as urban infrastructure, circular economy, renewable energy, regenerative farming and soil health, and the protection of freshwater and marine environments.

At this meeting, our keynote speaker is Prof. David MacMillan, who will talk about novel types of catalysis in chemical reactions, allowing chemistry to be performed using environmentally friendly approaches. Prof. MacMillan was awarded the Nobel Prize in chemistry last year for his work in this area. In the first main session, we will look at strategies to reduce carbon dioxide in the atmosphere, including the use of hydrogen to generate electricity and technologies to capture carbon dioxide from the atmosphere. Our second session will look at the manufacture of eco-friendly products through what is known as "green chemistry," and the pros and cons of plastics and plastic substitutes.

It is clear that science and technology are very valuable in addressing environmental challenges. However, they are not enough, because science and technology are mere tools that need to be used responsibly. The more fundamental need is to change human consciousness and attitude toward nature. Thus, our third main session will examine ways to engage the public, such as through effective policy making and ways to encourage the spirit of volunteerism.

The work of HJIFUS in restoring the environment has relied on the unique vision and inspiration provided by Rev. Dr. Sun Myung Moon, combined with the efforts of Rev. Dr. Hak Ja Han Moon, who worked side-by-side with him to substantialize that vision on a global scale. I am happy to inform you that their sixty-second wedding anniversary takes place on April 16, in just a few days. Please join with me in congratulating them on this occasion.

Another significant day this month is April 22, which is observed as Earth Day. I am pleased to announce that one year ago, HJIFUS launched an online, bimonthly publication, *The Earth & I*. Its purpose is to educate the general public on environmental issues, and to inspire people to take an active role in caring for the environment.

Finally, I wish to thank the iPeaceTV technical team and time zone coordinators for their tremendous investment in setting up and operating the virtual platform, which connects everyone together, as if we were in the same location. In addition, I am grateful for the support we have received from The Washington Times Foundation in the United States and *The Segye Times* in Korea.

I am looking forward to a very stimulating and productive meeting.

Thank you very much. ◊

About the Speaker

- Chair, Hyo Jeong International Foundation for the Unity of the Sciences
- Former Chair and President of *The Washington Times* and United Press International, and several other international news publications under News World Communications, as well as American Life TV Network and other electronic media
- Former President, World Media Association, The Washington Times Foundation, American Family Coalition, and Times Aviation International
- Former International President of the Youth Federation for World Peace
- Former Co-Chair of the Board of Trustees of the University of Bridgeport, Connecticut
- An honorary doctorate in Political Science from Sun Moon University; M.Ph. in Political Science, George Washington University

Opening Session
Congratulatory Address
Dr. Jeong Kee Hong
Vice-Minister of the Environment, Republic of Korea

Hello! My name is Jeong Kee Hong, Vice-Minister of the Environment in the Republic of Korea.

I would like to offer my wholehearted congratulations to the participants of the Twenty-Eighth International Conference on the Unity of the Sciences organized under the theme, "Investigating Pathways to Resolve Environmental Challenges." I think the conference is very timely, as we move beyond mere climate change and head into an era of climate crises.

I would like to extend my sincere gratitude to Dr. Sun Jin Moon, ICUS XXVIII Conference Chair Dr. Douglas Joo, Nobel laureate Dr. David MacMillan, and other distinguished scholars who are attending this renowned conference.

The big wave of the climate crisis and the COVID-19 pandemic are challenges humankind cannot avoid, and they must be addressed to safeguard our survival.

On April 5 this year, the IPCC announced that limiting global warming to around 1.5°C requires the reduction of global greenhouse gas emissions by 43% by the year 2030. The international community must listen carefully to this warning and raise high the flag of carbon neutrality. Through various initiatives, South Korea has also been striving to transform itself into a carbon-neutral society. Last year, the Republic of Korea confirmed its net-zero scenario and presented the nation's updated NDC (nationally determined contribution) target of 40% below the 2018 level by 2030 to the international community.

At the same time, South Korea is investing its efforts to reduce fine dust and transition to the era of low-carbon energy, such as coal phase-out, by supplying zero-emission vehicles such as electric and hydrogen vehicles. In addition, South Korea plans to limit single-use plastics and recycle plastics into materials that reduce the amount of waste. I believe that the government's efforts will expedite the nation's transition to a circular economy.

This year, the Ministry of Environment

will actively support the industry's green transition by applying the newly established K-Taxonomy as a pilot program. Efforts toward carbon neutrality should accompany innovation in society at large. A paradigm shift will also be required to move toward a carbon-neutral society.

I sincerely hope that today's gathering, in which leading scholars of the world are present from the fields of environment and energy, will serve as a platform to discuss climate change and the manufacture of eco-friendly products. I also hope it will help our society take one more step toward carbon neutrality.

Thank you. ◊

About the Speaker

- Eighteenth Vice-Minister of the Environment, Republic of Korea (March 2020–May 2022)
- Served as the Director of the Water Management Policy Office, Director of the Environmental Policy Office and Head of the reassessment on the Four Major Rivers restoration project
- PhD in Environmental Studies, Seoul National University, Republic of Korea

Opening Session
Founder's Address
Dr. Sun Jin Moon
Representing the Founders

Greetings!

It is an honor and privilege to be a part of this historic Twenty-Eighth International Conference on the Unity of the Sciences (ICUS XXVIII), which includes highly accomplished scientists and scholars from around the world, united in this great mission to aid, protect, preserve, heal, and restore all life on this precious planet.

It is my esteemed pleasure to welcome our honorable keynote speaker, Prof. David MacMillan, an eminent scientist and recipient of the 2021 Nobel Prize in Chemistry. I am also very pleased to welcome and thank all of the distinguished scientists, scholars, and guests participating in ICUS XXVIII. Furthermore, I am very grateful to the organizers for putting together this magnificent program.

I am Sun Jin Moon, daughter of Rev. Dr. Sun Myung Moon and Rev. Dr. Hak Ja Han Moon, founders of the ICUS series. I am delighted to welcome you and share my parents' insights and vision. We applaud your outstanding scientific and scholarly accomplishments, which have laid a new foundation that can contribute to restoring our natural environment.

The advances in human civilization are clearly discernible through the marvelous discoveries, innovations, and inventions that have benefited humankind. However, on the flip side, modern civilization has also produced fearsome weapons of war and pollution, and it has encouraged the gross consumption and unsustainable practices that now threaten life on Earth.

Although my parents are widely known as religious leaders, they have long recognized that resolution of the world's problems requires the efforts of the great minds of science coupled with values derived from the higher power of Divine consciousness and one's personal conscience. Thus, my parents founded the ICUS series in 1972, to find solutions to critical human problems through interdisciplinary dialogue, guided by absolute values—that is, values that benefit everyone, worldwide. Since then, ICUS meetings have been held in two phases. The first phase ended in the year 2000, when greater attention had to be given to more pressing issues of new conflicts in the world. Unfortunately, over the passing years, environmental issues grew in seriousness, with the increasing destruction of nature's pristine beauty and life-sustaining gifts. Concerned about these life-threatening problems, my beloved mother revived the ICUS series in 2017, focusing the meetings on addressing global environmental issues.

My mother has devoted her life to the service of humanity and the environment, working side-by-side with my father until his passing in 2012. Since then, she has continued to lead numerous peace organizations around the world. As a result, she has come to be recognized globally as the "Mother of Peace" and has inspired people worldwide to unite for peace and mutual prosperity, guided by universal values that serve all life on this planet and beyond.

Today we are in the throes of multiple existential crises everywhere. That is why the works you present here, we hope, can echo in the collective conscious minds of the global population to usher in a new age of peace and prosperity for all. In her memoir, *Mother of Peace*, my mother said[1]:

True peace certainly requires that we resolve the current conflicts between religions, races, and nations. The even greater challenges that we face, however, include the destruction of the environment and demographic trends. The world's leading peace awards focus on solving the problems of the present generation. Yet we must solve the problems of the present in a way that is integrated with a practical vison for a happy future... For decades, my husband and I took on the task of ensuring that humanity moves into a future assured of abundant food supplies and pleasant healthy environments.

Her revival of the ICUS series was based on the practice of unifying all resources to serve humanity and the planet, not just in theory but actual, real-world applications that can save the Earth. Therefore, it is my honor to address you all on behalf of my parents, who see your work to be a vital part of the foundational solutions to the compounding modern-day problems that besiege the planet's health.

Studies of human society and our natural environment show that we are all interconnected and interdependent, with every sector of our lives entangled in a virtual or actual web. Yet, in the world today, interdependence is being replaced with individual self-interest, leading to divisions in every arena, whether at school, home, work, or in society. We are confronted with global conflicts, extremism, acts of terror, violence, racism, hate, bullying, and daily crimes against our fellowman.

So, what is the solution to all this conflict and division in humanity that is destroy-

---

1 Hak Ja Han Moon, *Mother of Peace: And God Shall Wipe Away All Tears from Their Eyes* (Washington, DC: The Washington Times Global Media Group, 2020), 27–28. <https://motherofpeace.com>

ing all life? My mother answered this question at ICUS XXV, in 2019, as follows[2]:

> *We can take proper dominion or stewardship over the natural world only when we ourselves come to embody God's love. Through us, the natural world becomes connected to the true love of the intangible God, who is without form. For this reason, God, human beings, and the natural world are meant to form an organic whole, harmonized and united in true love, in accordance with God's purpose of creation. In this process, religion, which seeks God, and science, which deals with the world of material reality, are themselves harmonized and unified.*

Now, more than ever, it is incumbent upon us to "embody God's love" as the natural world is in an existential crisis. All life on Earth hangs in the balance, due to climate change brought about by human activity. Just this month, on April 4, 2022, the Intergovernmental Panel on Climate Change (IPCC) published the Sixth Assessment Report (AR6) contribution by Working Group III (WGIII), which includes this statement[3]:

> *The assessment of the Physical Science Basis (IPCC WGI AR6) documents sustained and widespread changes in the atmosphere, cryosphere, biosphere, and ocean, providing unequivocal evidence of a world that has warmed, associated with rising atmospheric CO_2 concentrations reaching levels not experienced in at least the last 2 million years. Aside from temperature, other clearly discernible, human-induced changes beyond natural variations include declines in Arctic Sea ice and glaciers, thawing of permafrost, and a strengthening of the global water cycle (WG1 SPM A.2, B.3 and B.4). Oceanic changes include rising sea level, acidification, deoxygenation, and changing salinity. Over land, in recent decades, both frequency and severity have increased for hot extremes but decreased for cold extremes; intensification of heavy precipitation is observed in parallel with a decrease in available water in dry seasons, along with an increased occurrence of weather conditions that promote wildfires.*

Already in 2018, the IPCC warned that with global warming likely to increase by 1.5°C or more, it will have disastrous effects on all life around the globe.[4] With extreme disruptions of weather, and greater and more frequent risks of floods, drought, fires, extreme storms, and weather events, all living things are at risk. The rise in sea level is a risk to coastal

---

2 Hak Ja Han Moon, Founder's Address, *Proceedings of the Twenty-Fifth International Conference on the Unity of the Sciences.* (Washington, DC, USA, and Gapyeong-gun, South Korea: Hyo Jeong International Foundation for the Unity of the Sciences Press, 2019), 8.

3 Working Group III Contribution to the IPCC Sixth Assessment Report, "Climate Change 2022: Mitigation of Climate Change" (AR6 WGIII Final Draft Chapter 01), 1–9. <https://www.ipcc.ch/report/ar6/wg3/>

communities, increases risks to agriculture and infrastructure, leading to economic hardships and subsequent migration.

Along with the mass extinction of species, loss of ecosystems and coral reefs, and the biodiversity needed to sustain life on Earth, the IPCC stresses the need for action and sustainable change. The comprehensive IPCC reports and their collective data on climate change finds our future to be catastrophic for all life if we do not get carbon emissions down to the net-zero range by 2050. It is a jarring wake-up call for the world to stand in solidarity and to take action for the sake of all living creatures on this planet, with no time to waste.

Global survival and welfare are threatened at every moment, and that is why at this conference, we have a great responsibility and opportunity to offer solutions to the environmental challenges facing humanity. Life as we know it hangs in the balance of our choices and actions. The interdependent fate of humanity and the Earth is a direct result of not knowing who we are and why and how we are living. We need to have the sacred wisdom to know that all life on Earth is connected to the heart of Divine love of the highest power. When science can embody God's love, a peaceful world of enlightenment and heart can be passed on to generations of life to flourish and thrive. If there is no shift of conscience or consciousness, then the current state of this Earth will be bleak.

In his autobiography, *As a Peace-Loving Global Citizen*, my father illuminated the path we should take[5]:

> *The best environmental movement, however, is the one that spreads love. People generally take care of things that belong to them or to people they love. They do not, however, take care of or love the natural environment that God created. God gave this environment to humanity. It is His will that we use the environment to obtain food, to have it in abundance, and to experience the joy of living in the beauty of nature. Nature is not something to be used once and thrown away. Our descendants for many generations to come must be able to rely on it just as we have. The shortcut to protecting nature is to develop a heart that loves nature.... To care for and love the environment is to love God.*

The Earth is our one and only home. It is the very fabric of our existence. We are all indebted to planet Earth for our survival. Earth enables reproduction and

---

4 V. Masson-Delmotte et al. (eds.), "Summary for Policymakers," *IPCC Special Report: Global Warming of 1.5 °C* (Cambridge, UK, and New York, USA: Cambridge University Press, 2018), 3–24. <https://doi.org/10.1017/9781009157940.001> <https://www.ipcc.ch/sr15/chapter/spm/>

5 Reverend Sun Myung Moon, *As a Peace-Loving Global Citizen* (Washington, DC: The Washington Times Foundation, 2009), 311. English translation from the original Korean publication by Gimm-Young Publishers, Seoul, Korea.

nurtures us as a mother's womb. The lives we live are because humanity is one family under God. We should no longer regard the environment as a commodity or tool to be used at our whim. We must cherish and value all life as our family, regarding the people of the world as our global beloved brothers and sisters, and caring for all creation as we raise our gardens and farms, or sharing unconditional love with our precious family pets. We must embody God's love, then the natural world will reciprocate that love a thousandfold. We should respect and protect the Earth as one vital, living organism, not with a "me" but a "we" mentality. Each of us can take initiative to that effect on the individual level, then to the family, tribe, nation, and world.

Gathered here today are leaders capable of solving environmental problems through science and technology. I hope this conference, anchored in a sense of responsibility and embodying God's heart, will prove a valuable opportunity for each of you to share luminous ideas and research results. I look forward to hearing your presentations and deliberations during this conference.

Thank you very much. ◊

About the Speaker

- Fifth daughter of Rev. Dr. Sun Myung Moon and Rev. Dr. Hak Ja Han Moon
- Senior Vice President, Women's Federation for World Peace International
- Chair, Pacific Rim Education Foundation in Kailua-Kona, Hawaii (since 2012)
- Has served the Family Federation for World Peace and Unification, Universal Peace Federation, and Women's Federation for World Peace (since 2000)
- Baccalaureate in psychology from Harvard University
- Inducted into the Alpha Sigma Lambda National Honor Society
- Master of Arts degree in Psychology from Columbia University
- Awarded an honorary doctorate in social science from Sun Moon University

Opening Session | Keynote Address
New Catalytic Strategies for a Sustainable Future
Professor David W.C. MacMillan
Chemistry, Princeton University, USA

Introduction

Catalysis impacts nearly every aspect of the modern world (**Figure 1**). Today, at least 90% of industrial-scale chemical reactions employ catalysis, and fully 35% of the world's gross domestic product is based on catalytic processes. Industrial-scale catalytic processes provide many of modern society's staples, including food, medicines, solar cells, diagnostic tools, and even polymers and materials.

Over the next century, catalysis will provide solutions to many pressing societal challenges, such as alternative energy, environmental remediation, inexpensive pharmaceuticals, sustainable agriculture, and renewable soft materials. As such, the development of efficient, environmentally friendly catalysts will be key to our common goal of creating a more sustainable future.

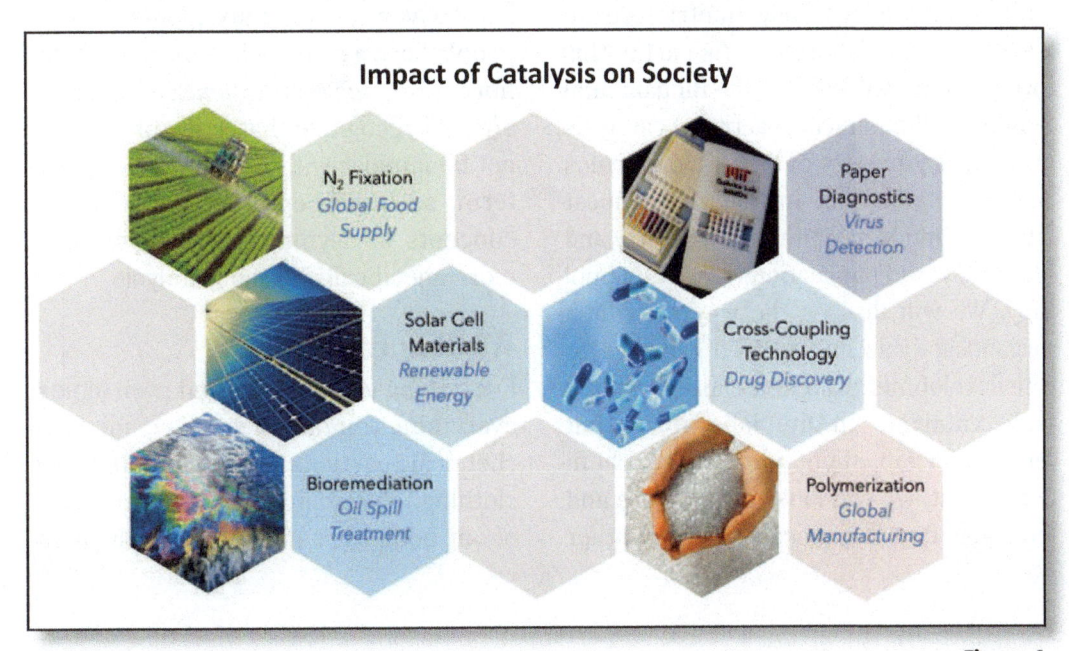

Figure 1

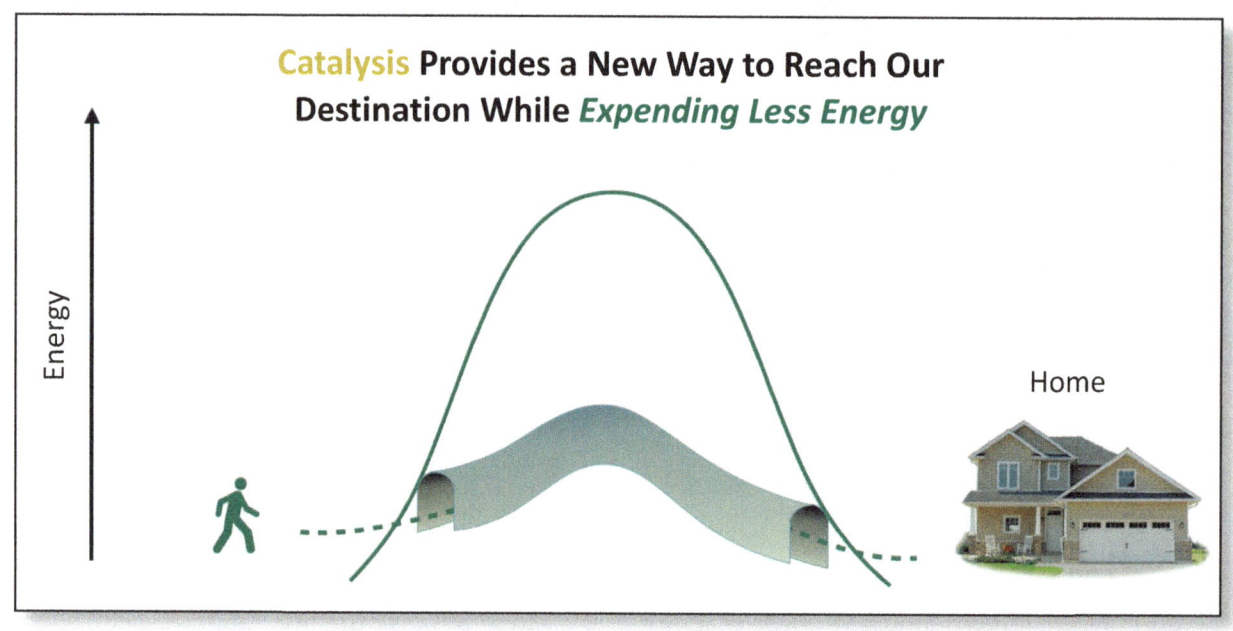

Figure 2. Catalysis Provides a New Way to Reach Our Destination While *Expending Less Energy*

Over the past two decades, my research group has been dedicated to inventing powerful and environmentally friendly catalytic processes. In this lecture, I will provide a brief overview of our efforts to contribute toward a sustainable world. We will first discuss the advent and development of asymmetric organocatalysis in my laboratory. As part of this overview, we will explore the fundamental concepts of chemical reactivity, catalysis, and the asymmetry of organic molecules. We will examine the impact of organocatalysis on modern synthetic chemistry and see real-world applications of this technology. We will discuss the ways in which organocatalysis has created a bridge to the development of a new class of sustainable catalysts in my laboratory that employs visible-light photoredox chemistry. Last, we will look to the future and consider how organocatalysis and other sustainable approaches—including photocatalysis, electrocatalysis, and biocatalysis—will continue to impact scientific research and society as a whole.

It was my great honor to share the 2021 Nobel Prize in Chemistry for the development of asymmetric organocatalysis. Over the past six months, many people have asked me just one basic question: *What exactly is asymmetric organocatalysis?* I recognize that many of you may not be familiar with this rather technical term, so I will begin by explaining the concepts of asymmetry, catalysis, and more specifically, organocatalysis.

What is catalysis?

Everything you can see—and even almost everything you *cannot* see—is made from chemical reactions. Our food, medicines, clothing, electronics, and even the cells of our bodies are created through chemi-

The Importance of Catalysis: Industrial Nitrogen Fixation

$$N_2 + 3\,H_2 \xrightarrow{Fe/Ru\ cat} 2\,NH_3$$

~50% of nitrogen atoms in our bodies come from synthetic ammonia

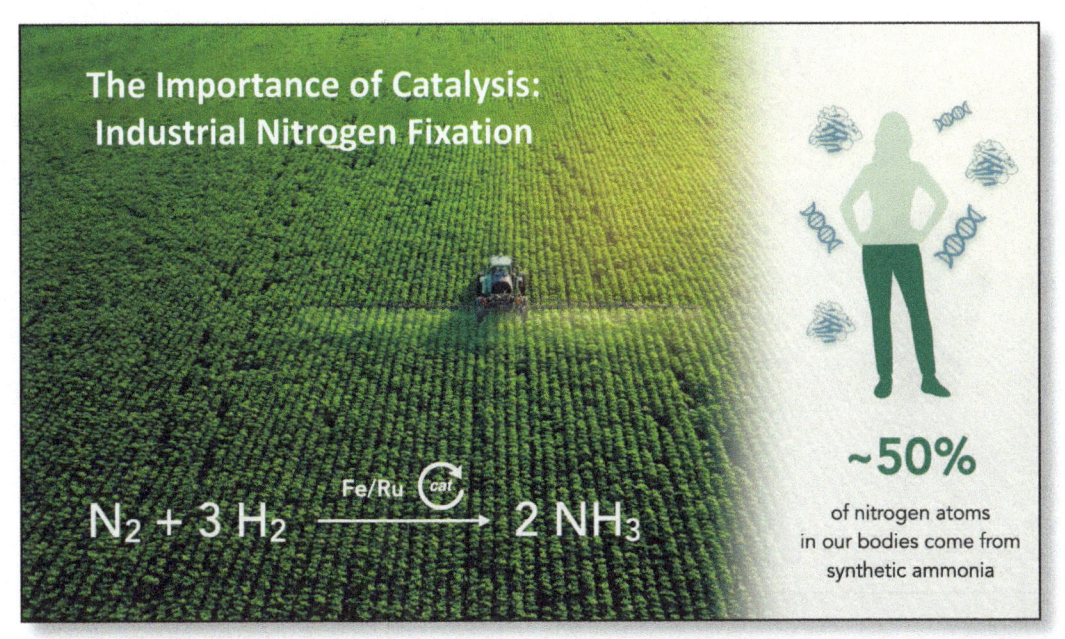

Figure 3

cal reactions. We can appreciate that chemical reactions are absolutely essential to our existence. But how do chemical reactions work? It turns out that chemical reactions usually do not happen spontaneously: In order to convert a starting material into the product you desire, you often need to input a significant amount of energy. We can use the process of catalysis to dramatically lower the amount of energy needed to make a chemical reaction proceed.

Here is an analogy I use to explain the power of catalysis to my students: Imagine that every evening, you have to climb a large hill to get from class back to your dorm. Your daily trek up and over this hill clearly requires a lot of energy (**Figure 2**). Now, imagine that, one day, you find a tunnel that leads directly *through* the hill. You walk through the tunnel and straight home with hardly any effort. In this analogy, catalysis represents the tunnel, as it provides a new path for chemical reactions to follow with minimal input of energy. An important feature of catalysts is that they are not used up in chemical reactions. In fact, a very small amount of a potent catalyst can be used to convert large quantities of starting materials to products. Catalysis therefore allows chemical reactions to proceed quickly, efficiently, and often with relatively little cost.

These unique attributes help explain why so much of our modern society relies on catalysis. As an example, we can consider the real-world impact of perhaps the most revolutionary catalytic process of our age: *nitrogen fixation* (**Figure 3**). Invented in the early twentieth century by the German chemist Fritz Haber, catalytic nitrogen fixation is an industrial-scale process that converts nitrogen (N_2) to

What About Asymmetric?

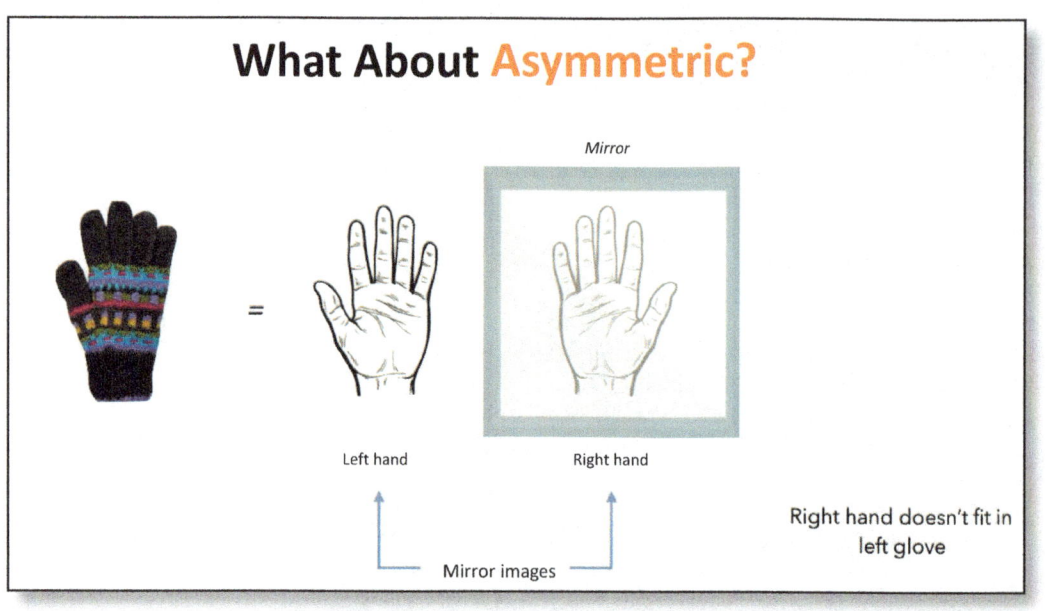

Figure 4

ammonia (NH$_3$). Ammonia is an essential component of farming, and modern agriculture depends on the ammonia produced through catalytic nitrogen fixation to grow an abundance of food crops. This increased crop productivity, in turn, has fueled the exponential growth of Earth's population over the past century. Without this one catalytic process, it would be very difficult—if not impossible—to produce the food needed to sustain Earth's 8 billion people. To visualize how this process impacts you, consider that about half of the nitrogen atoms in your body were likely derived from industrial-scale nitrogen fixation.

What is asymmetry?

The property of asymmetry is easy to visualize (**Figure 4**). Most of us have two hands, and our hands are mirror images of one another. These mirror images are *almost* identical, but they are not *exactly* identical since your hands are not superimposable. That is, your right-handed glove does not fit on your left hand. Two mirror-image objects, like hands and feet, that are almost identical but not superimposable are called *asymmetric*.

Interestingly, molecules can also exhibit this property of asymmetry. In other words, organic molecules can exist in one of two nonsuperimposable mirror image forms, each of which is called an *enantiomer*. As you might expect, the two mirror images of an organic molecule have almost identical physical properties and behavior, and it can be difficult to distinguish one from the other in a laboratory. In fact, two enantiomers will behave in very similar ways to one another *except* when they are interacting with other asymmetric molecules. Much like your two mirror-image hands can dis-

How Can We Distinguish between Mirror Images?

(R)-carvone

(S)-carvone

"A whiff"

(R) Smells like spearmint

(S) Smells like caraway

Figure 5

tinguish between two mirror-image gloves, each enantiomer of an organic molecule will interact differently with other asymmetric molecules.

Here is an example that may be familiar to those of you who took organic chemistry in college: (R)-carvone and (S)-carvone are enantiomers of one another. Since the carvone mirror images are almost identical, it is very difficult to distinguish (R)-carvone from (S)-carvone in the lab unless you have access to fairly specialized and expensive equipment. However, if you were to hand a sample of each carvone enantiomer to a very small child, she would immediately be able to differentiate between the two mirror images. Why is that? The reason is that, to us, (R)-carvone smells like spearmint and (S)-carvone smells of caraway (**Figure 5**).

Why is it that our smell receptors can so readily perform a task that is so challenging for the most sophisticated lab equipment? The answer is that the human body is made of building blocks (proteins, DNA, carbohydrates, and hormones) that themselves exist as asymmetric molecules. Therefore, just as your left-hand glove interacts differently with your left and right hands, your asymmetric smell receptors interact differently with each carvone enantiomer, recognizing different and easily distinguishable scents.

This property of asymmetry that infuses your body has important implications beyond the actions of your smell receptors. Perhaps most impactful is the role of asymmetry in determining how medicines behave in the body (**Figure 6**). Many medicines are small organic molecules that can exist as two mirror images. Not surprisingly, your body typically responds differently to each of these mirror images. One of these mirror images

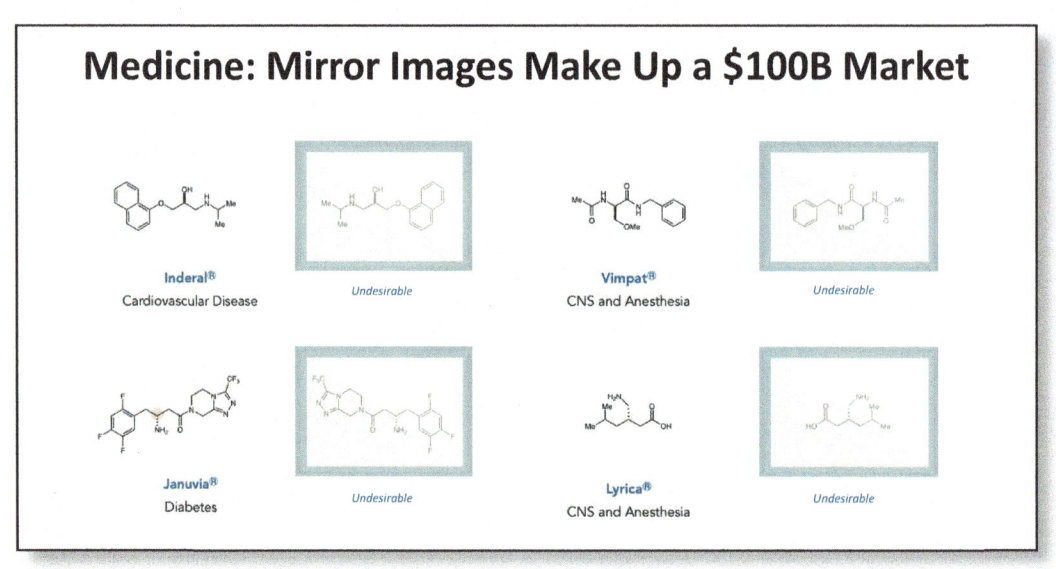

Figure 6

will interact with your body in the desired way, perhaps by blocking the activity of a problematic overactive enzyme. However, the other mirror image will likely not play a productive role and might even interact with your body in dangerous ways. It is easy to see why we need technologies that allow us to selectively prepare *only* the desired mirror image of a medicinal compound. A long-standing priority of organic chemistry is to invent new strategies that employ catalysis to construct molecules in single–mirror image form. This field of research is known as *asymmetric catalysis.*

What is organocatalysis?

Today, organocatalysis represents a major field of asymmetric catalysis, and I am proud to have played a role in pioneering this exciting, sustainable new technology.

To place organocatalysis into historical context, we must consider the state of the field of asymmetric catalysis at the dawn of my career, over two decades ago. In 1996, there were two major modes of asymmetric catalysis: *biocatalysis* and *metal catalysis* (**Figure 7**). Biocatalysis employs naturally occurring enzymes to construct single enantiomers of organic molecules. This strategy leverages the fact that enzymes are themselves massive asymmetric molecules. Metal-based asymmetric catalysis, on the other hand, is a man-made field that uses catalysts composed of metals paired with single-enantiomer organic molecules.

Here is where my part in this story begins. On finishing my PhD studies under the great Prof. Larry Overman at the University of California at Irvine, it was my privilege to undertake postdoctoral research with Prof. David Evans at Harvard University. Dave is one of the most influential chemists in the world and an absolute master of asymmetric metal catalysis, and my time in the group provided me invaluable exposure to this field of research.

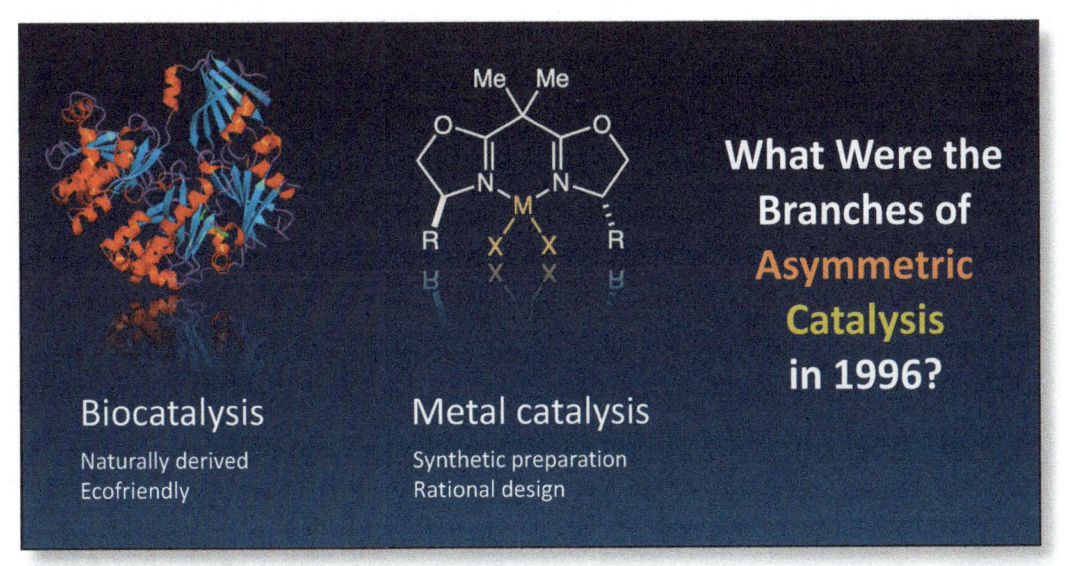

Figure 7

While in the Evans group, I grew to fully appreciate both the power and the drawbacks of asymmetric metal catalysis. The reality was that most of my days were spent working in a rather uncomfortable and bulky contraption called a *glovebox* (**Figure 8**). A glovebox is a special piece of equipment designed to rigorously eliminate moisture, oxygen, and air from a chemical reaction. In the Evans group, we needed to work in the glovebox, as the metal component of the asymmetric metal catalyst can be difficult to handle. Often, metals cannot be exposed to the atmosphere; moreover, they can be toxic, unsustainable, and expensive. Organic molecules, on the other hand, are usually quite easy to handle and are typically safe, sustainable, recyclable, and inexpensive.

During my time in the group, I began to wonder whether it would be possible to develop a new kind of asymmetric catalyst

Figure 8

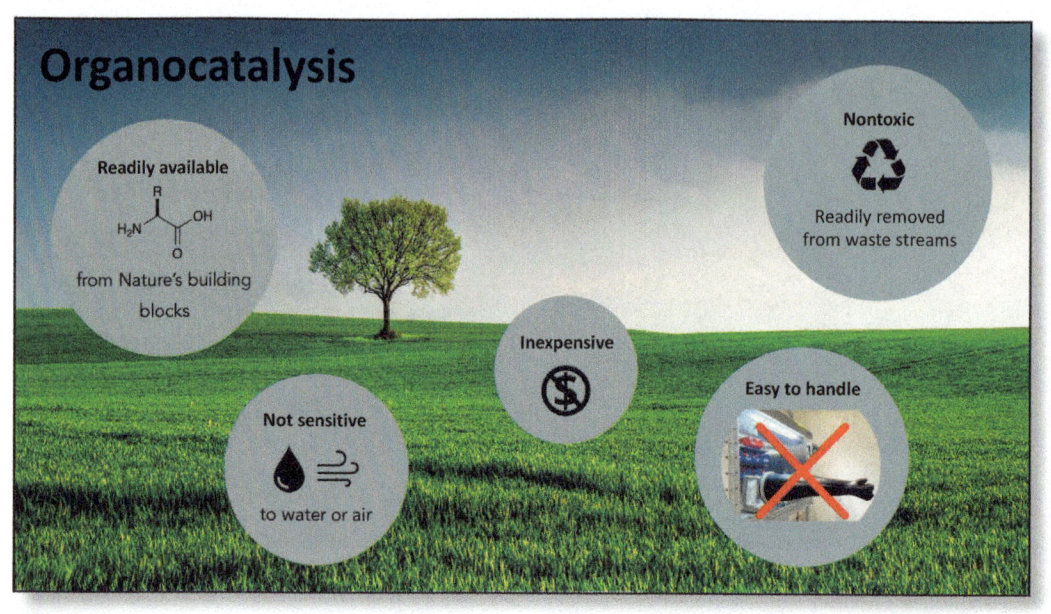

Figure 9

that *only* used the single-enantiomer organic component and eliminated the metal altogether. I believed that an asymmetric catalyst based solely on organic molecules could have revolutionary potential, for several reasons. Organic catalysts should be easily and inexpensively constructed from Nature's building blocks. Organic catalysts would not be sensitive to moisture or air and could be handled without special equipment. Additionally, organic molecules are sustainable, recyclable, and nontoxic. Finally—and most exciting to me— I recognized the possibility of developing a single general platform for organic catalysis that could be used to promote not just one but hundreds of different reactions. I imagined that, if I could achieve this goal, then perhaps one day organic molecule catalysis could ultimately mature into a third major pillar of asymmetric catalysis.

As I was formulating this grand vision, I unfortunately had no idea of exactly *how* we might leverage simple organic molecules as asymmetric catalysts. I could not know at the time that within two years of the start of my independent career at the University of California at Berkeley, we would invent a general platform for asymmetric catalysis that completely eliminated the metal component. In our seminal 2000 publication, we gave this new field of asymmetric catalysis a name: *organocatalysis* (**Figure 9**).

Discovery of organocatalysis

When I arrived as an assistant professor at the University of California at Berkeley in the summer of 1998, I really did not know how I would accomplish my goal of developing a general method for organocatalysis, but I had faith in the stellar, devoted, and incredibly hardworking group of young graduate students who

An Amine-Catalyzed Diels-Alder Reaction

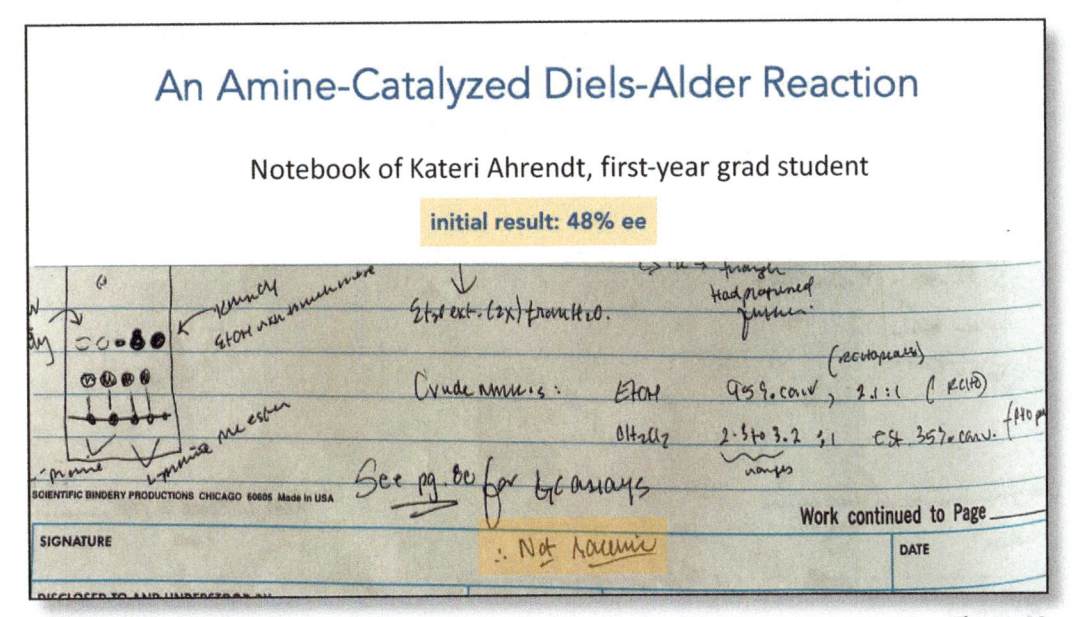

Notebook of Kateri Ahrendt, first-year grad student

initial result: 48% ee

Figure 10

joined my lab that first year. Fortunately, my confidence in my team was well placed.

In the spring of 1999, a first-year graduate student in my group, Kateri Ahrendt, found that a small organic molecule was capable of catalyzing a well-known reaction called the *Diels-Alder cycloaddition*. Most excitingly, as Kateri wrote in her notebook, the reaction was *not racemic:* the organocatalyst was able to *preferentially* form the desired mirror image of the product (**Figure 10**). In the absence of this catalyst, the reaction generates both mirror images of the product in equal quantities. Although this preliminary result was far from publication-ready—we would ultimately need to modify the structure of the catalyst and optimize reaction conditions in order to achieve really useful levels of selectivity—Kateri's pivotal experiment of April 3, 1999, represented the first demonstration in my lab of an asymmetric organocatalytic reaction.

Following a great deal of experimentation, we ultimately hit upon a highly effective, generalizable organocatalyst scaffold: the *imidazolidinones* (**Figure 11**). From a sustainability standpoint, the imidazolidinones are really desirable catalysts since they can be made easily and inexpensively by combining phenylalanine, an amino acid, with acetone, a bulk chemical commonly used as a paint stripper.

Imidazolidinones also have the important advantage of being highly tunable; that is, their structures can be easily modified to meet the particular needs of different types of chemical reactions. This tunability would allow us to ultimately realize the grand vision of developing a *generic activation mode:* a single organocatalyst scaffold that could be applied to hundreds of different chemical reactions. In fact, the

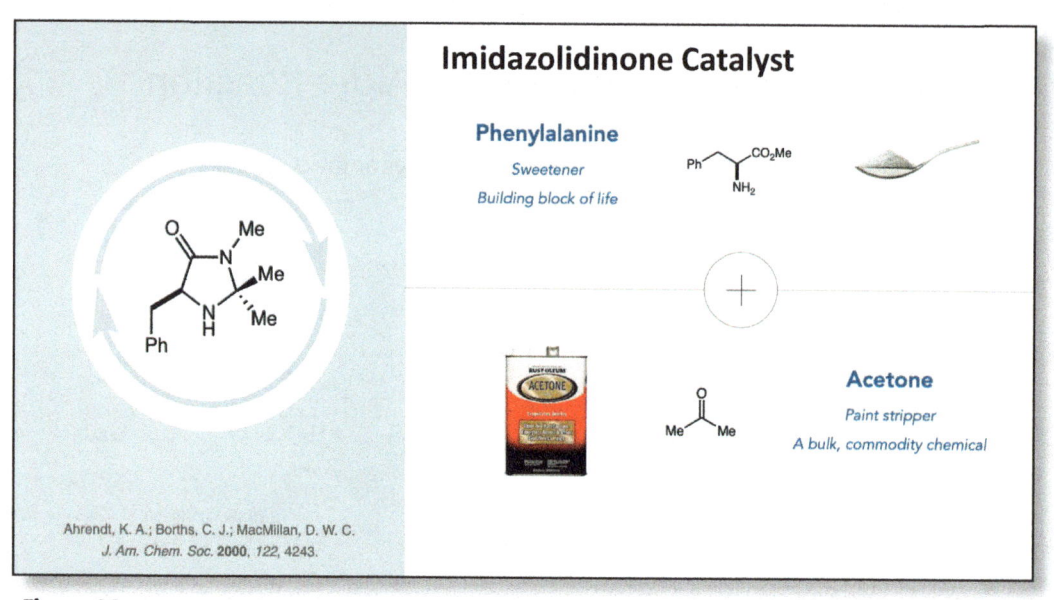

Figure 11

emergence of organocatalysis as a major mode of asymmetric catalysis can be traced to this key catalyst design feature.

Following our landmark 2000 publication, in which we reported the first asymmetric organocatalytic Diels-Alder cycloaddition, we went on to develop a series of asymmetric organocatalytic reactions using the imidazolidinone scaffold. A second-generation imidazolidinone catalyst, brilliantly engineered by graduate students Joel Austin and Chris Borths, proved even more versatile than our original scaffold, allowing us to quickly develop dozens of powerful new asymmetric organocatalytic reactions.

Expansion of organocatalysis

Around this time, other academic researchers began to make important contributions to the growth of this new field, most notably Karl Anker Jørgensen and Yujiro Hayashi (**Figure 12**). Meanwhile, Ben List and Carlos Barbas were conducting elegant research in the related area of enamine-based organocatalysis.

This was an incredibly exciting time, as our group and others around the world were inspired to invent a wide swath of powerful new reactions that made use of the asymmetric organocatalysis framework. Of course, all transformational scientific advances are built upon the foundations of their forebears, and the field of asymmetric organocatalysis is highly indebted to the many outstanding chemists who have made fundamental contributions in adjacent areas of catalysis. Without the discoveries of these pioneers, the field of asymmetric organocatalysis simply could not exist.

As the field of asymmetric organocatalysis continued to grow, we also began to branch out in exciting new directions. Of particular interest to our group, from a sustainability standpoint, was the pos-

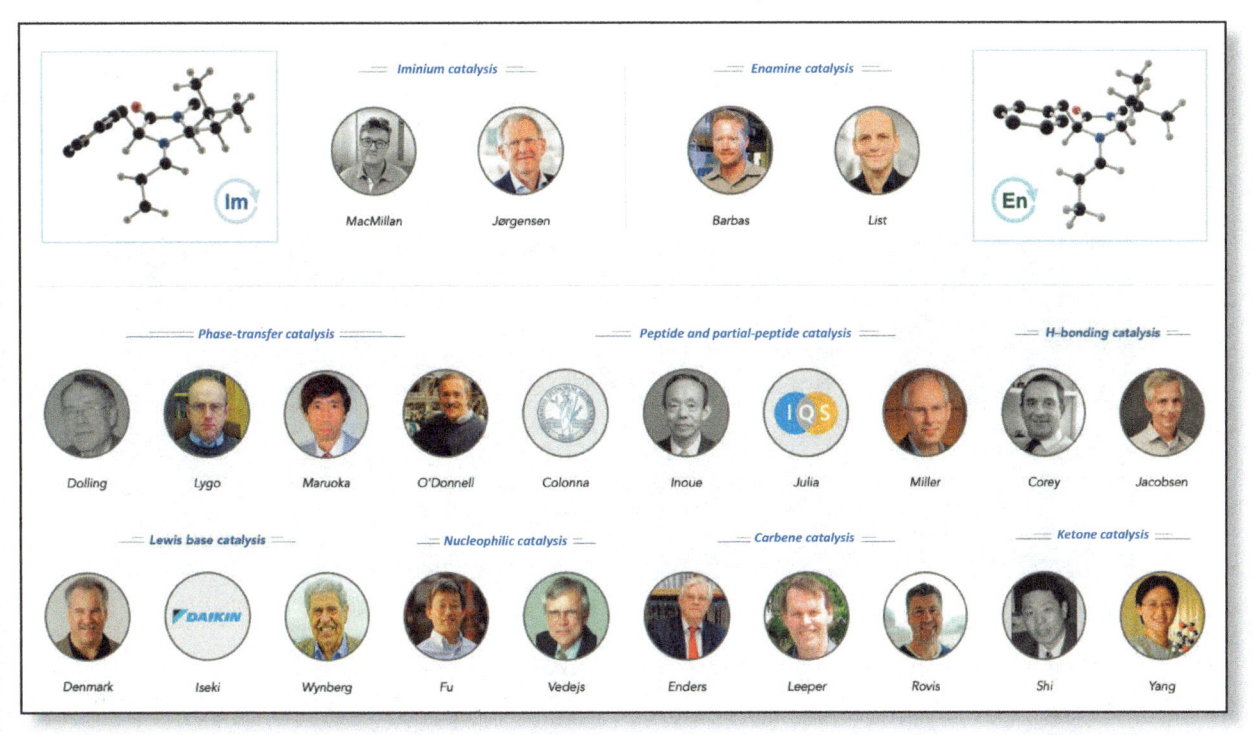

Figure 12

sibility of merging multiple organocatalytic reactions together within a single reaction vessel as a way to quickly—and with minimal waste—build up a high degree of chemical complexity from simple starting materials. This general strategy, which we termed *cascade cataly-* *sis,* would actually emulate the way that Nature makes complex molecules. In Nature, simple building blocks are shunted through a biochemical assembly line wherein each enzyme catalyzes a distinct reaction in a controlled sequence to quickly generate complex end products.

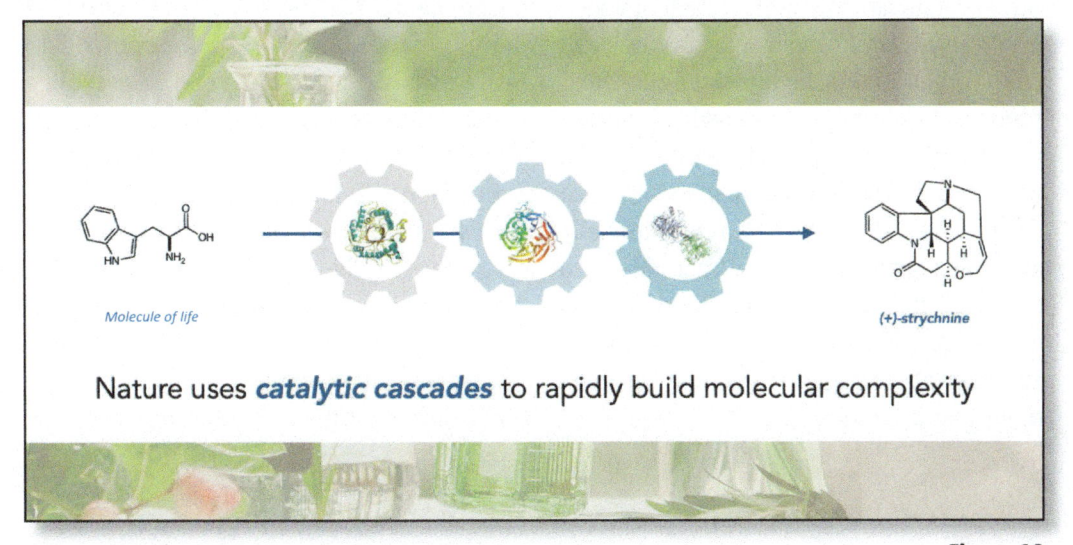

Figure 13

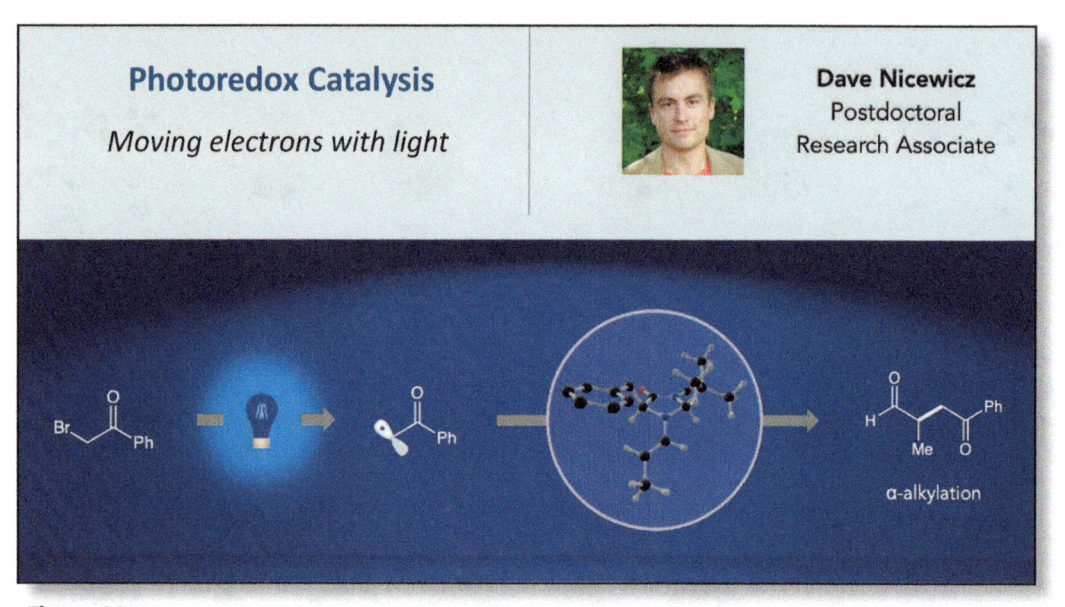

Figure 14

Our analogous cascade catalysis strategy, which used simple organocatalysts in place of Nature's enzymes, proved highly effective. In a key demonstration, we accomplished a rapid total synthesis of strychnine, a naturally occurring molecule that is also a commonly used rat poison (**Figure 13**). The central complexity-building transformation was accomplished in a single reaction vessel, as a very simple starting material was fed through three consecutive organocatalytic cycles, each of which added an element of complexity to the molecule, to generate a highly elaborated end product. This product was easily converted to strychnine, allowing us to achieve a rapid synthesis of this challenging natural product in just 12 steps from commercially available starting materials. Cascade organocatalysis has since been further validated as a sustainable, waste-efficient, and highly economical strategy for building complex molecular architectures.

Photocatalysis

In 2007, Teresa Beeson, an outstanding third-year graduate student in my lab, developed a novel mode of asymmetric organocatalysis, which we termed *SOMO catalysis,* that would ultimately launch our research group into some really exciting new directions, culminating in the development of a new type of sustainable catalytic platform that combines organocatalysis with visible-light catalysis. This new area, called *photoredox catalysis,* was first demonstrated by an excellent postdoctoral researcher in my group, Dave Nicewicz (**Figure 14**). The ability to merge organocatalysis with visible-light catalysis represented an extremely important advance, and over the past 14 years, photoredox catalysis has matured into an important field of research in its own right. In fact, today, the field of photoredox catalysis is as influential as the field of organocatalysis, and I feel very for-

Figure 15. Organocatalysis for a Circular, Recyclable Plastics Economy. "PET" plastics → Organic catalyst → Degradation of plastic waste. Stanford University, IBM. Bob Waymouth, James Hedrick.

tunate to have been deeply involved in the conceptualization and advancement of both of these crucial areas.

Organocatalysis and society

I am proud of the ways in which asymmetric organocatalysis has influenced the field of synthetic organic chemistry over the past 20 years. The impacts of organocatalysis can also be felt beyond the confines of the academic research community. In industrial settings, where environmentally responsible practices are emerging as a major corporate priority, organocatalytic processes are particularly appealing, as they are sustainable and remove the need to employ costly, toxic, and non-renewable metals. As such, organocatalytic solutions are increasingly applied to modern, large-scale industrial processes.

Today, bulk-scale organocatalysis is used in the environmentally friendly synthesis of scented fragrances, particularly those manufactured by the Swiss company Firmenich. Organocatalysis has also found application in the recyclable plastics economy. For example, Prof. Bob Waymouth of Stanford University and Dr. James Hedrick of IBM have developed organocatalytic processes that break down polymers to their component monomeric building blocks (**Figure 15**). Since these monomers can then be transformed back to polymers, such organocatalytic processes have the potential to render plastics completely recyclable and sustainable. Needless to say, the widespread adoption of such technologies would have an enormous impact on our oceans and other threatened ecosystems.

Perhaps not surprisingly, asymmetric organocatalysis has been heavily adopted across the pharmaceutical industry, where the need to access single-mirror-image versions of medicinal molecules is paramount. Merck's chronic migraine

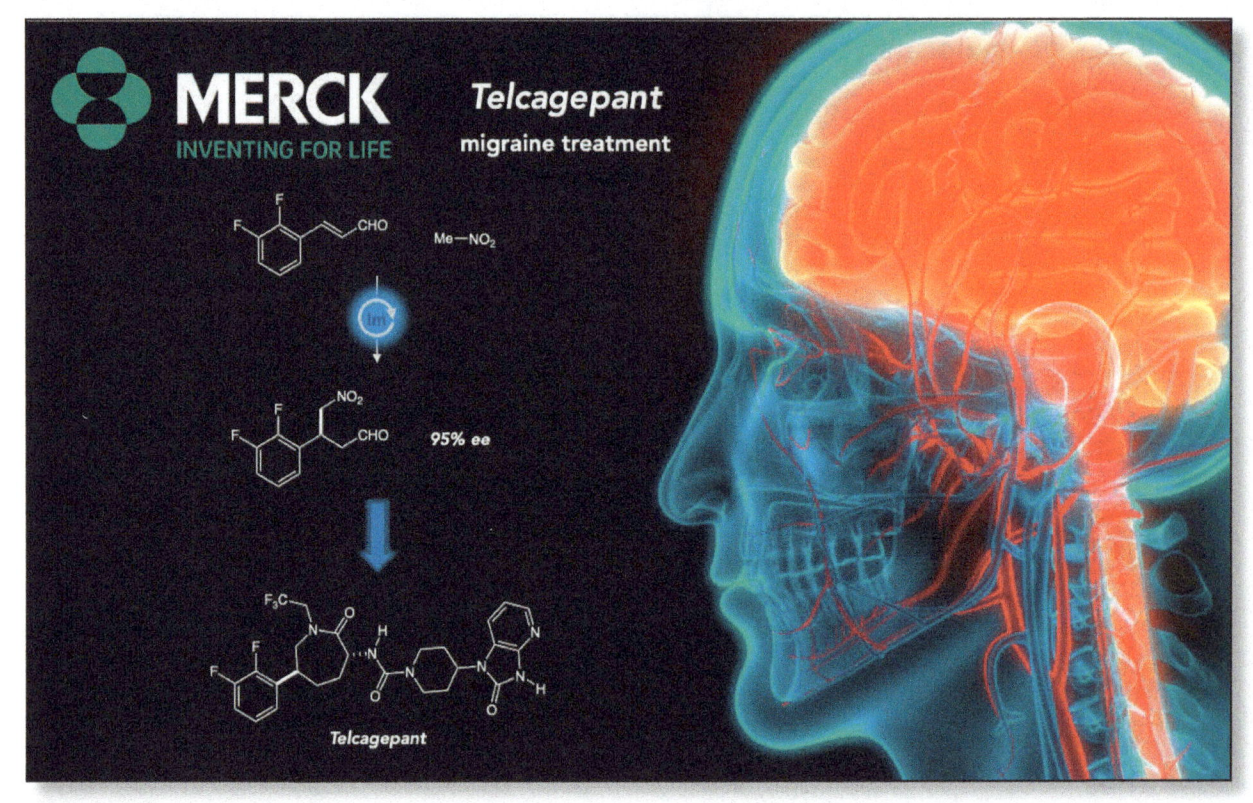

Figure 16

drug, Telcagepant, for example, is manufactured using asymmetric organocatalysis techniques developed in our laboratory (**Figure 16**).

Beyond industrial applications, organocatalysis has influenced our broader society in somewhat surprising ways. It turns out that organocatalysis has played an important role in democratizing the field of chemistry (**Figure 17**). Organocatalysts are inexpensive, and organocatalytic reactions can be carried out under atmosphere without special equipment. For that reason, organocatalysts are uniquely accessible to scientists and educators around the world. Across the globe, students and researchers have the unique opportunity to gain hands-on experience in cutting-edge asymmetric organocatalysis technologies and, perhaps more importantly, to make their own innovative contributions to this field of research, regardless of the financial and instrumental resources available to them. The accessibility and ease of use of organocatalysis stand in stark contrast to many other modern synthetic methods, and the implications of this democratizing effect are exciting to consider. I would argue that the next revolutionary advances in organocatalysis will emerge not from the most well-resourced labs but from those researchers who have the best and most creative ideas.

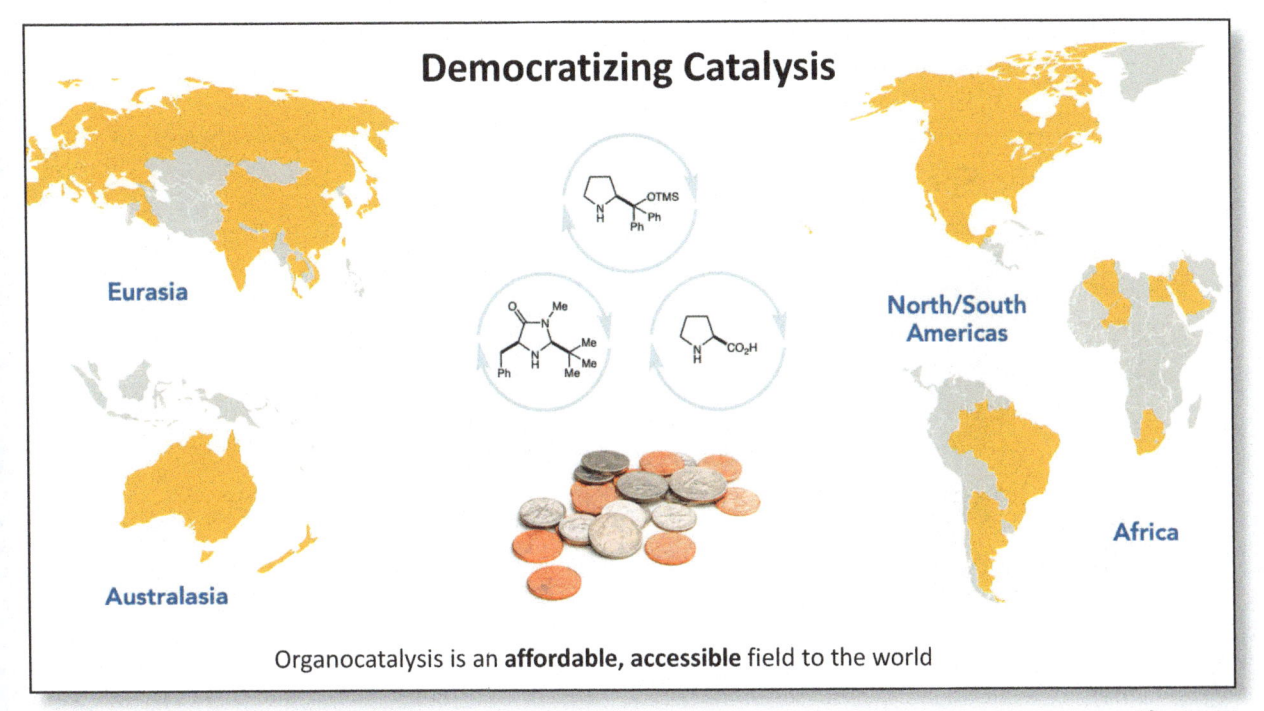

Figure 17. Organocatalysis is an **affordable, accessible** field to the world

The future of catalysis

I am often asked what the future holds for organocatalysis. I do not have an answer to that question, but I know that we must provide for our expanding global population in environmentally responsible ways. I believe that the solutions to many of our most pressing challenges will depend upon scientists' ability to develop powerful and sustainable catalyst systems (**Figure 18**).

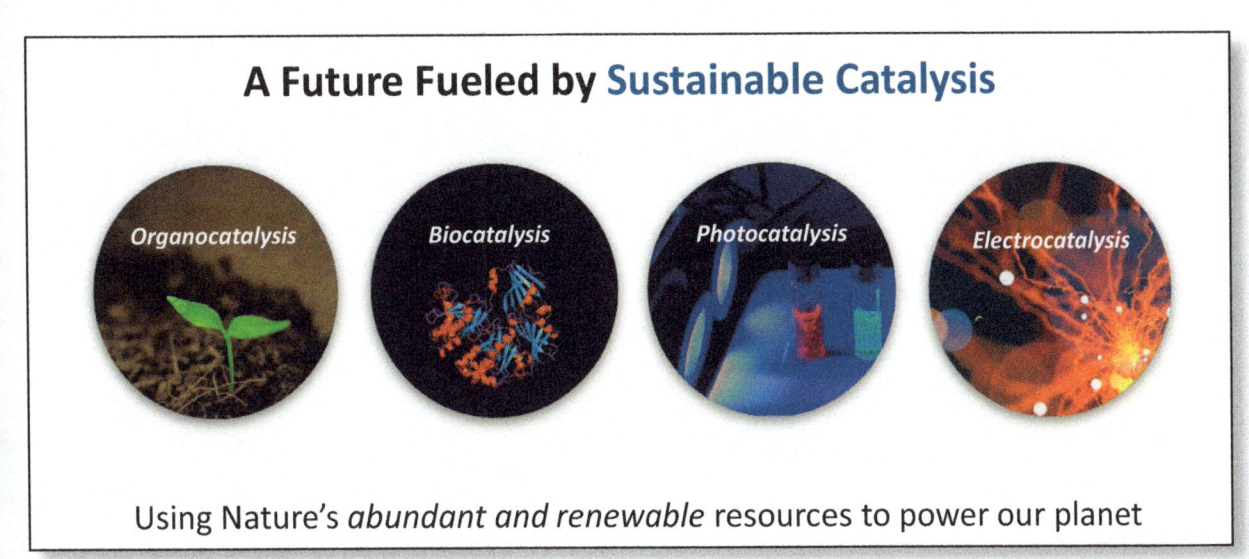

Figure 18. A Future Fueled by Sustainable Catalysis — Organocatalysis, Biocatalysis, Photocatalysis, Electrocatalysis. Using Nature's *abundant and renewable* resources to power our planet

These solutions will include organocatalysis and biocatalysis, but they will also include emergent sustainable technologies, such as photocatalysis and electrocatalysis. ◊

About the Speaker

- James S. McDonnell Distinguished University Professor of Chemistry, Princeton University, USA.
- Researcher in organic synthesis and catalysis, including photoredox, SOMO, enamine, and iminium catalysis, and coined the term *organocatalysis*.
- Corecipient of the Nobel Prize in Chemistry (2021) for his work in asymmetric organocatalysis.
- Elected member, National Academy of Sciences, USA.
- Editor in Chief, *Chemical Science* (2010–2015).
- PhD in Chemistry, University of California, Irvine, USA.

ICUS XXVIII

SESSION 1

Addressing Climate Change: Strategies to Achieve "Net Zero"

About Session 1

Addressing Climate Change: Strategies to Achieve "Net Zero"

Climate change can be mitigated by achieving "net zero"—a term that refers to attaining a balance between the amount of greenhouse gases emitted and the amount removed from the atmosphere. This session presents two types of strategies to reach the "net zero" goal.

One strategy is to replace the use of fossil fuels with "green" hydrogen—that is, hydrogen produced by the electrolysis of water using a renewable energy source such as wind or solar power. The production and combustion of green hydrogen avoids the production of carbon dioxide. The first presentation explains the use of hydrogen to power an array of vehicles, and to provide heat and electricity in houses.

The second presentation shows how "net zero" can be achieved using several approaches that reduce the amount of carbon dioxide in the atmosphere. One approach is to use biological measures, involving the ability of plants to capture carbon dioxide through photosynthesis, produce carbon-based compounds, and support ecosystems. Another approach is to build facilities for bioenergy with carbon capture and storage. In a third way, called "direct air capture," carbon dioxide is removed from the air through chemical reactions and the carbon is stored in various forms, such as carbonate rocks. ◊

About the Chair, Cliff I. Davidson

- Thomas C. and Colleen L. Wilmot Professor of Engineering; Environmental Engineering Program Director in the Department of Civil and Environmental Engineering; Center of Excellence in Environmental and Energy Systems, Syracuse University, New York, USA.
- Founding Director, Center for Sustainable Engineering.
- Research interests include sustainable development in urban areas, transport and fate of pollutants, sources of airborne particles, and assessment of green infrastructure for stormwater management.
- Fellow, American Society of Civil Engineers.
- Fellow, Association of Environmental Engineering and Science Professors.
- Fellow, American Association for Aerosol Research.
- Awards include: United Methodist University Scholar-Teacher Award (2014); William H. and Frances M. Ryan Award of Carnegie Mellon University for Meritorious Teaching (2009); AEESP Outstanding Educator Award (2007).

Session 1 | Presentation One

Tapping Hydrogen for Electric Power and Transportation

Professor David Blekhman
Sustainable Energy and Transportation
California State University Los Angeles, California, USA
Technical Director, Hydrogen Research and Fueling Facility

Thank you for the opportunity to share with you the exciting developments around hydrogen and fuel cell technologies. Indeed, as discussed later, there are significant investments planned for these technologies, both in the US and the rest of the world. These investments are motivated by a combination of climate change concerns and attempts to restart the economy after the COVID-19 shock.

In addition, the hydrogen economy is far more universal than the economy based on lithium batteries, where few countries and companies have access to the mining and processing of lithium into battery electrode materials. In contrast, and as examples that follow will show, many countries and many industries can economically benefit from working in the hydrogen space. And the latter is growing from individual deployments to bigger concepts like *hydrogen hubs* or *hydrogen valleys,* which both refer to conglomerations of industrial and research organizations in a geographical area aimed at devel-

About the Speaker

- Technical Director, California State Los Angeles Hydrogen Research and Fueling Facility.
- Professor of Technology, California State University, Los Angeles, USA.
- Fulbright Distinguished Chair in Alternative Energy Technology, Chalmers University, Sweden (2019–2020).
- Cal State LA Outstanding Professor Award (2021–2022).
- Expert in hydrogen infrastructure, alternative and renewable technologies, fuel cells, and hybrid and electric vehicles.
- Contributor, *Forbes*.
- PhD in Mechanical Engineering, University at Buffalo, USA.

oping initiatives for the efficient production, storage, transport, distribution, and use of hydrogen.

Several examples that follow will introduce you to the California State University Los Angeles (Cal State LA) campus, which already serves as a hydrogen hub. The Cal State LA hub includes such sustainable technologies as a solar photovoltaic installation, robust electric vehicle (EV) charging, a fleet of fuel cell vehicles, and the Hydrogen Research and Fueling Facility. As we proceed to consider similar combinations, the concept will continue to repeat in various forms and sizes.

Fuel cells are electrochemical generators that take fuel and an oxidizer to produce electricity. Their advantage is that, unlike combustion engines, conversion to electricity is direct and at a higher efficiency than in their thermomechanical counterparts. The majority of such devices discussed today are called *proton exchange membrane* (PEM) fuel cells, utilizing hydrogen for the fuel and oxygen from air as the oxidizer. Conveniently, in addition to producing electricity, the only emissions are pure water and heat, which are appropriately managed. While the history of hydrogen and fuel cells is 180 years old, the more recent practical application was implemented for space exploration in the Apollo program (1966) and then in the Space Shuttle program (alkaline fuel cells), as shown in **Figure 1**. The advantage of fuel cells was also apparent for the same reasons they have benefit now: For the long haul, they are a much lighter

First Practical Applications: Space

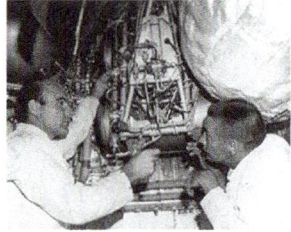

PEM fuel cells being installed in a Gemini 7 spacecraft (Source: Smithsonian Institution, from the Science Service Historical Images Collection, courtesy of General Electric)

1.5 kW for Apollo missions

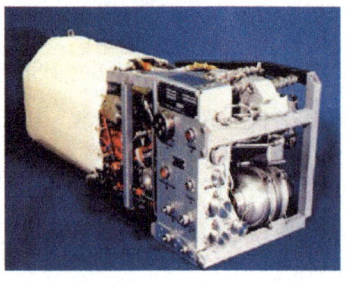

Each fuel cell power plant is 14" high, 15" wide and 40" long and weighs 255 lbs
The V~A range of each is 2 kW at 32.5 Vdc, 61.5 A, to 12 kW peak at 27.5 Vdc, 436 A.
7 kW maximum continuous power.
Total power ~21 kW with 15-min peaks of 36 kW.
The orbiter power consumption ~ 14 kW.
Service between flights until 2000 h.

The alkaline fuel cell system as used on the Space Shuttles.
Three modules per shuttle

http://www.doitpoms.ac.uk/tlplib/fuel-cells/history.php

Figure 1

system than batteries. The Space Shuttle program utilized three 7 kW fuel cells in each shuttle. Systems up to 10 kW can be used in homes, forklifts, or portable generators. New systems have grown to 100 or 200 kW per fuel cell stack and can be chained to any power required.

The Cal State LA Hydrogen Research and Fueling Facility is the world's largest campus-based on-site generation station. It uses electricity and water to produce 60 kg/day of hydrogen, which is sufficient to fuel up to 20 passenger vehicles per day. It was funded in 2008–2009, with active construction taking place in 2010–2011. After additional upgrades, we celebrated its grand opening in May 2014 and have been providing motorists with hydrogen ever since. While operating as a fueling facility, the goal is to conduct applied research, workforce training, and public outreach. Over the years, the facility has hosted more than 10,000 visitors, where 85% were K–12 and college students and the rest were industry and government professionals.

We also enable many other projects through custom fueling and technical assistance. We are approaching close to US$1.5 million in additional funding to the university from various agencies enabling these programs. The latest is US$500,000 awarded in February 2022 from the California Energy Commission for the zero-emission vehicle (ZEV) workforce curriculum focusing on our internship program.

In November 2014, the Cal State LA facility became the first in the world to

Cal State LA Hydrogen Research and Fueling Facility Specs

Production: 60 kg/day
Storage: 60 kg
Pressure: 350 and 700 bar
Capacity: 15–20 fuel cell vehicles per day

Figure 2

demonstrate accurate metering for the sale of hydrogen by the kilogram directly to fueling customers. The facility made the first retail sale by the kilogram to Volkswagen and Audi while the vehicles were test-driven at the 2014 LA Auto Show. This achievement had a further national impact, as the National Institute of Standards and Technology adopted new hydrogen meter accuracy requirements informed by this experience.

The station deploys a Hydrogenics electrolyzer, first- and second-stage compressors capable of fast-filling at 350 and 700 bar, 60 kg of hydrogen storage (see the three tanks on the left in **Figure 2**), 10 kg of high-pressure buffer storage, water purification, and various cooling systems to operate at −20 °C. The station is powered by the campus grid, which includes a solar power plant that produces renewable electricity during the day. The station is built in full compliance with Society of Automotive Engineers standards J2600, J2601, and J2719.

Equipped with state-of-the-art data collection hardware for all of its systems, the facility actively collaborated with the National Renewable Energy Laboratory (NREL) on collecting hydrogen fueling performance and operations data in 2014–2018. The information provided was incorporated into the NREL's aggregate reporting. These reports explicitly acknowledged Cal State LA as a contributor.

The Hydrogenics electrolyzer the station uses is a model HySTAT-A 1000, encapsulated in a 20 ft container and

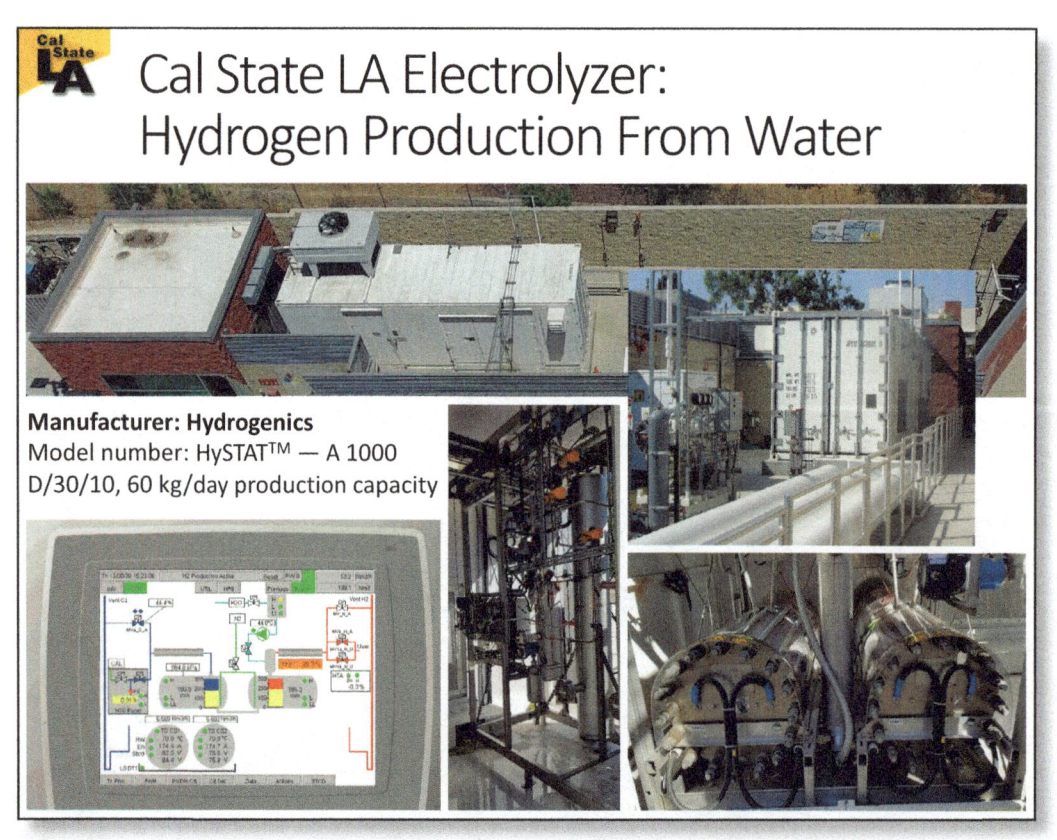

Figure 3

capable of producing 60 kg/day, with an energy consumption of approximately 65 kWh/kg H_2, as shown in **Figure 3**. It takes 9 kg of reverse-osmosis water to produce 1 kg of hydrogen; 8 kg of oxygen is vented. In the reverse reaction in fuel cells, the process restores 9 kg of water from 1 kg of hydrogen. The Hydrogenics electrolyzer uses alkaline electrolyte, just as in the Space Shuttle fuel cell, instead of relying on a PEM. The electrolyzer has two fuel cell stacks. The produced hydrogen is processed in the dryers to remove any water vapor, with final purities reaching up to 99.999%.

In recent years, Hydrogenics was acquired by Cummins, allowing this traditional engine manufacturer to rapidly enter both pathways for hydrogen with fuel cells and electrolyzers. Overall in the industry, there is a trend for companies to acquire technology that is missing in their portfolios rather than develop their own. For example, Plug Power, a fuel cell manufacturer, acquired Giner, a promising electrolyzer developer. This allowed Plug Power to control production for locations where its fuel cells are deployed—for example, in warehouses with hydrogen forklifts.

One of the examples of operational research is presented in **Figure 4** and addresses the work completed by one of the high school students who volunteered her summer after eleventh grade to learn about hydrogen. We asked her to summarize our maintenance and operations records of failure and repair performed on various subsystems. We were inter-

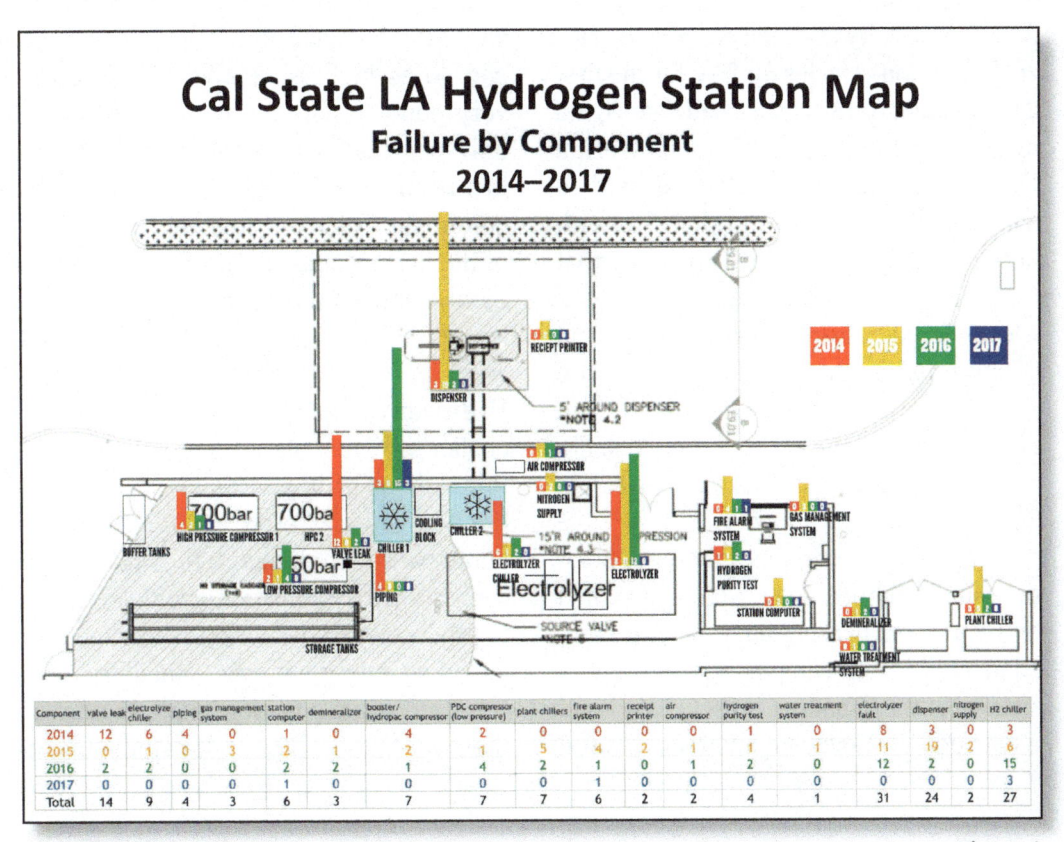

Figure 4

ested in a time lapse of what failed and when. As you can see in **Figure 4**, some equipment was failing more during the first or second year of active use—so-called teething issues—while other components started failing later, indicating wear and tear or aging problems. In addition, this student created a fantastic 3-min video introducing the Cal State LA Hydrogen Research and Fueling Facility. It can be found on YouTube at https://youtu.be/PwvbFC1MDkw. Even more information and history is on our website (https://www.calstatela.edu/ecst/h2station).

In February 2019, Cal State LA unveiled a fleet of zero-emission fuel cell Hyundai Tucson vehicles for use by the university community in an urban, shared-mobility model. Before COVID-19, students and faculty used the service to drive zero-emission vehicles for free for the first 2 h. The fleet of 18 vehicles had 6 deployed in serving Parking and Transportation Department needs and 12 dedicated to shared mobility service. In the aftermath of the pandemic, the Waive company that managed the program did not survive. Eleven vehicles were returned to the manufacturer while the remaining seven continue to serve campus needs. We are planning to restart some of the shared mobility activities with fuel cell vehicles later in 2022 with another provider.

Commissioned in the summer of 2020, the campus 900 kW solar system, installed on the newest parking infrastructure and connected to the campus microgrid, consists of 2800 Canadian Solar 370 W modules and 19 Yaskawa inverters, rated 1036 kW and 902 kW respectively. The design provides the secondary function of shading the vehicles parked underneath. The flatter, carport design favors production of electricity during summer. This installation provides ample renewable energy for hydrogen generation and EV charging needs. In **Figure 5**, you can also see data on the solar power generated.

The campuswide EV charging network consists of 55 Level 2 and 6 fast chargers, all by Charge Point, and forms one of the most robust facilities of its kind. The Charge Point management portal is extremely well designed, with access to multiple data metrics. Currently, this is the largest EV charging infrastructure among 23 campuses in the California State University system. Level 2 chargers typically can furnish up to 7 kW per plug, while the fast chargers deployed with us can go as high as 50 kW, making them very popular. The network started with only two modern chargers in 2011 and has been growing due to the university community demand and the leadership of the Parking and Transportation Department, supported by faculty.

One of the most exciting opportunities working with students has been the Hydrogen Student Design Contest, which challenged students around the world to think about hydrogen technologies. The last time it was offered, in 2017–2018, students were challenged to "develop a design for a system that uses electricity to produce hydrogen for cross-market uses, including energy storage, ancillary services, and transportation fuel." As part of their offerings, Cal State LA students sub-

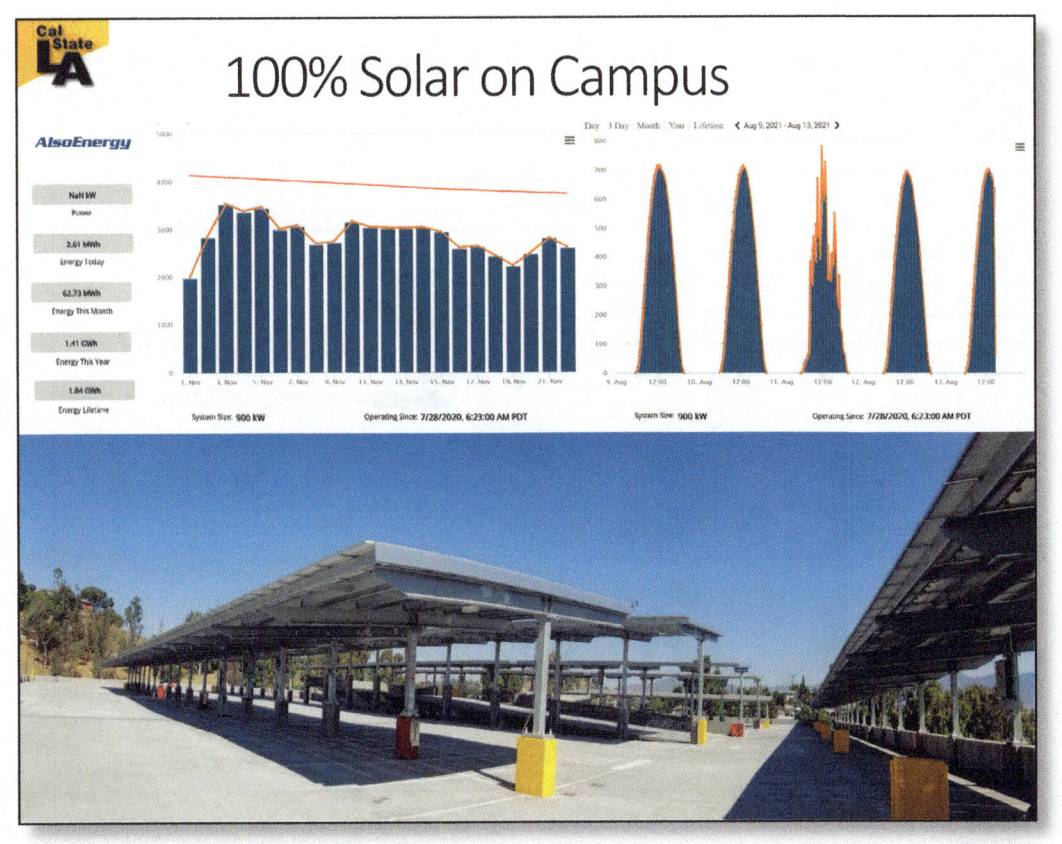

Figure 5

mitted the diagram in **Figure 6**, which reflects their vision for the integrated economy. They created a full representation of a hydrogen hub or valley in California, exceeding the assignment's expectations to some degree—though not surprisingly, because Cal State LA students get a lot of exposure to the hydrogen economy.

In our discussions for the hydrogen train used as a switch locomotive by BNSF Railway Company at its switchyard in the city of Commerce, California, one of the students suggested that tugboats would fall under the same model in being refueled in port. I was impressed with this forward thinking by our students. Since then, the concept of marine hydrogen has grown greatly, as discussed later.

The diagram (**Figure 6**) reflects the concept of generating renewable energy that is converted to hydrogen as a universal fuel that can be deployed in industrial applications like steelmaking and power production, and in commercial buildings (malls, schools, universities, office buildings, hospitals, and so on) and homes. As a transportation fuel, hydrogen can power passenger vehicles, buses, trucks, trains, ships, planes, and others.

In many ways, the Cal State LA campus has already approached the true definition of the hydrogen hub by encompass-

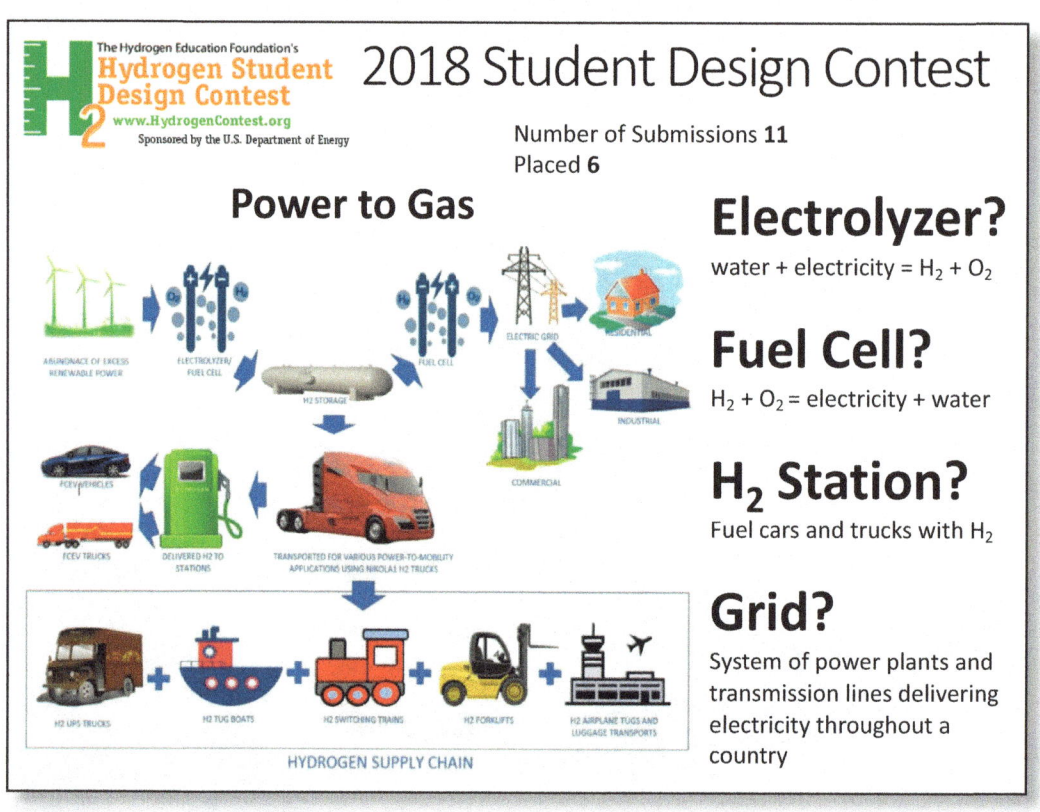

Figure 6

ing renewable energy generation, hydrogen production, fueling infrastructure, a fleet of vehicles, and jobs through its workforce development.

During my 2019–2020 Fulbright appointment as Distinguished Chair in Alternative Energy at Chalmers University in Gothenburg, Sweden, I was introduced to Hans-Olof Nilsson and Martina Wettin, the people behind creating residential and commercial hydrogen technologies in Sweden. One of their applications is the Nilsson house, where Hans-Olof himself lives. This is a unique creation, embodying the motto, "Yes, I can." The Nilsson house is a man's dream of modern, luxury living, made independent of the grid by deploying hydrogen technology. There are just a handful of such buildings in the world. More amazingly, it was completely conceived and executed by a single middle-class engineer and business visionary. The house incorporates most of the possible sustainable solutions, with the majority of them produced in Scandinavia and other European nations.

As shown in **Figure 7**, It starts with its energy for the year being produced by the photovoltaic panels on the roof and walls. The energy is stored in batteries for daily cycling, and hydrogen is stored seasonally for off-solar periods via the electrolyzer (made in Denmark). When energy and heating are needed for a Swedish winter, a 5 kW fuel cell by PowerCell, produced locally in Gothenburg, delivers both. Additionally, a geothermal heat pump assists with the heating and cooling needs of the house. Altogether, it is a remarkable set of technologies made to work for a grid-independent house. Under the brand name RE8760, the company is currently engaged in a multitude of projects throughout Sweden.

I was so impressed with this accomplishment that I felt compelled to thoroughly document the house as a case study for my future students. I raised about US$10,000 to produce a video that would introduce my students to the house and set them off on a learning journey about the dwelling's technology and about future directions or adaptations to other climate needs—for example, in California. The 15-minute documentary can be found on YouTube at https://youtu.be/j2Qpv1qz-2s.

California has been the trendsetter in clean transportation around the world, compelling the mighty and powerful automotive companies to deploy emission-free vehicles. Currently, California boasts the largest fleet of hydrogen-powered vehicles—in excess of 10,000 cars on the road. While this is not a large number, it is a start that allows us to resolve the everlasting chicken-or-the-egg conundrum. Original equipment manufacturers pro-

Figure 7

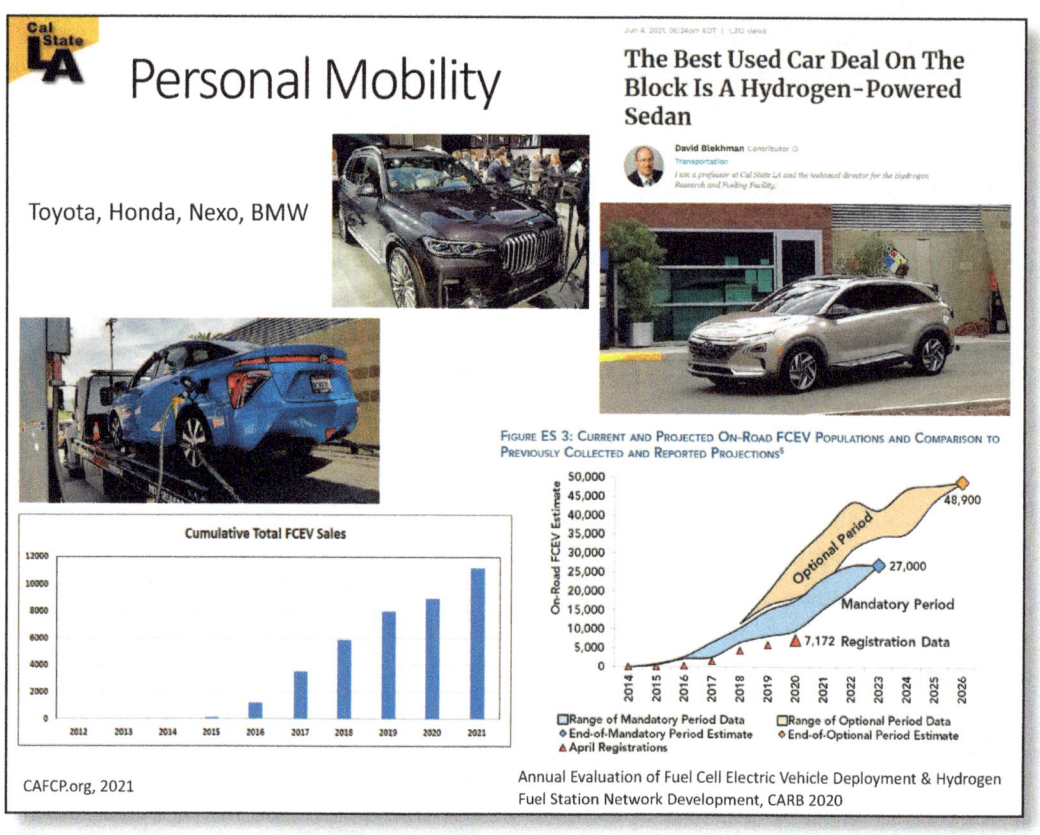

Figure 8

vide vehicles, and the state takes responsibility for the hydrogen infrastructure. Perhaps, as indicated on the graph in **Figure 8**, we might see up to 50,000 fuel cell electric vehicles (FCEVs) in a few years.

In addition, Toyota Mirai vehicles introduced several years ago are now coming back from their three-year leases and are sold through dealerships again at very affordable prices so that people from the lower-income bracket can also acquire zero-emission vehicles. In **Figure 8**, a Mirai is being refueled on the truck bed before transferring to its new owner after the initial 3-year lease. We saw dozens of these cars being refueled at Cal State LA during presale, inspiring me to write an article in *Forbes*: "The Best Used Car Deal on the Block Is a Hydrogen-Powered Sedan."

Honda, Hyundai, and BMW are also present in the US market. Fuel cells developed for passenger cars are now being deployed in many other applications.

California has the greatest number of operating public hydrogen stations in the US—49, with 124 more on the way. The map in **Figure 9** shows stations in Southern California, with the green ones already in operation. The development process was initially proposed as a cluster model, where several stations would be nearby and supporting each other during downtime. The

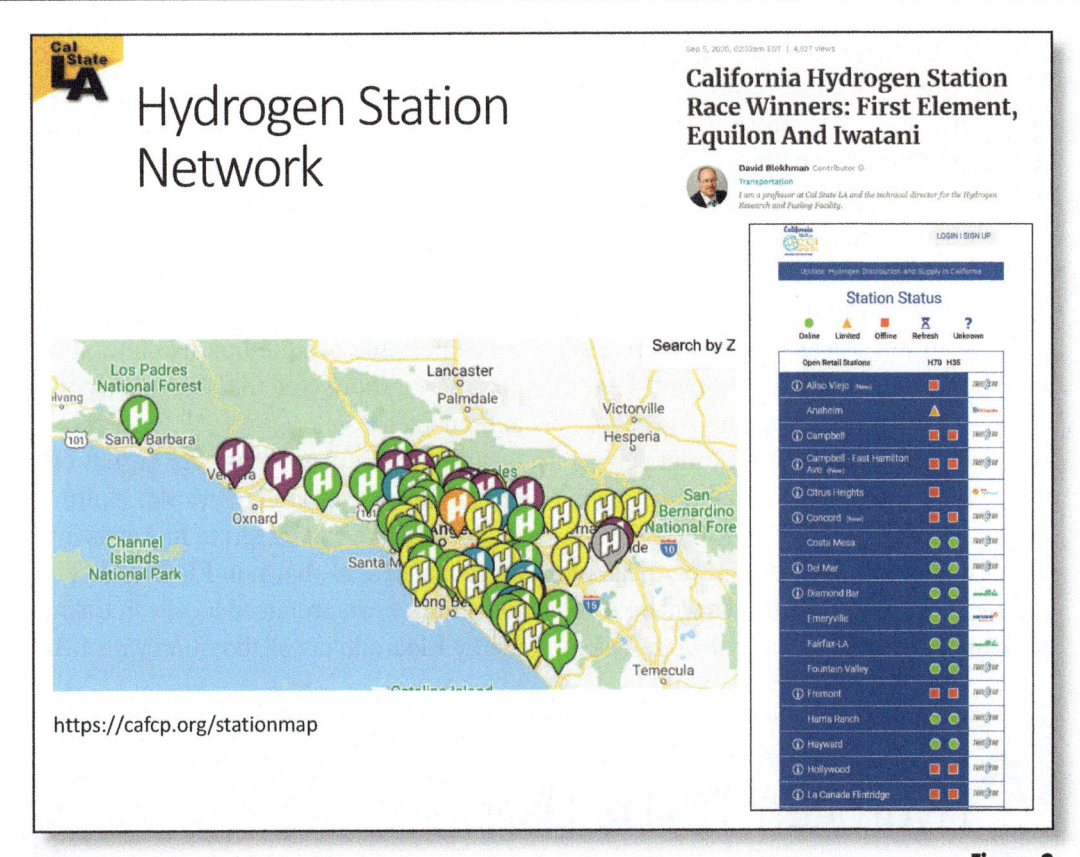

Figure 9

newer installations emphasize major corridors, with larger stations over 1000 kg/day retail and smaller backup stations with about 400 kg/day. In 2020, California granted awards for a more than 120-station network to be built by three companies in a few years. This would substantially move California toward its ultimate goal of 200 stations in several years. More details can be found in my *Forbes* article titled "California Hydrogen Station Race Winners: First Element, Equilon and Iwatani."

Of course, the hydrogen infrastructure has room to improve, because hydrogen supply sometimes has interruptions or stations might experience maintenance issues. To help drivers navigate station and fuel availability, the station-status SOSS website was developed, as depicted on the right side in **Figure 9**. Stations that are not currently fueling show in red, with some simply closed for the night due to local regulations.

Some companies are trying to develop their hydrogen network independently with private funding, especially for heavy-duty transport. The new concept being developed for their networks is called hub and spoke, where the hub produces hydrogen, fuels trucks and other vehicles on location, and delivers hydrogen to nearby stations.

In a report that is now three years old, the Institute of Transportation Studies at the University of California Davis conducted and evaluated potential penetration of ZEVs in the US market, as in **Figure 10**. For long-haul driving, hydrogen trucks have a big advantage over their battery counterparts, and the projection shows that in this category 80% of trucks will be powered by hydrogen. This might move even faster, because in 2020 California Gov. Gavin Newsom issued an executive order setting new statewide goals for phasing out gasoline-powered cars and trucks in California. Under the order, 100% of in-state sales of new passenger cars and trucks will be zero-emission by 2035; 100% of in-state sales of medium- and heavy-duty trucks and buses are to be zero-emission by 2045. According to the California Fuel Cell Partnership report, "Achieving California's Zero-Emission Future for Freight Movement," 70,000 hydrogen heavy-duty long-haul trucks will be needed to sustain the West Coast trade traffic. These trucks would serve not only in-state but also interstate commerce connected to California's ports and businesses.

There are now thousands of fuel cell buses deployed around the world. Multiple manufacturers can integrate fuel cells into their coaches, as shown in **Figure 11**. The two buses were produced locally in California by Eldorado in collaboration with BAE

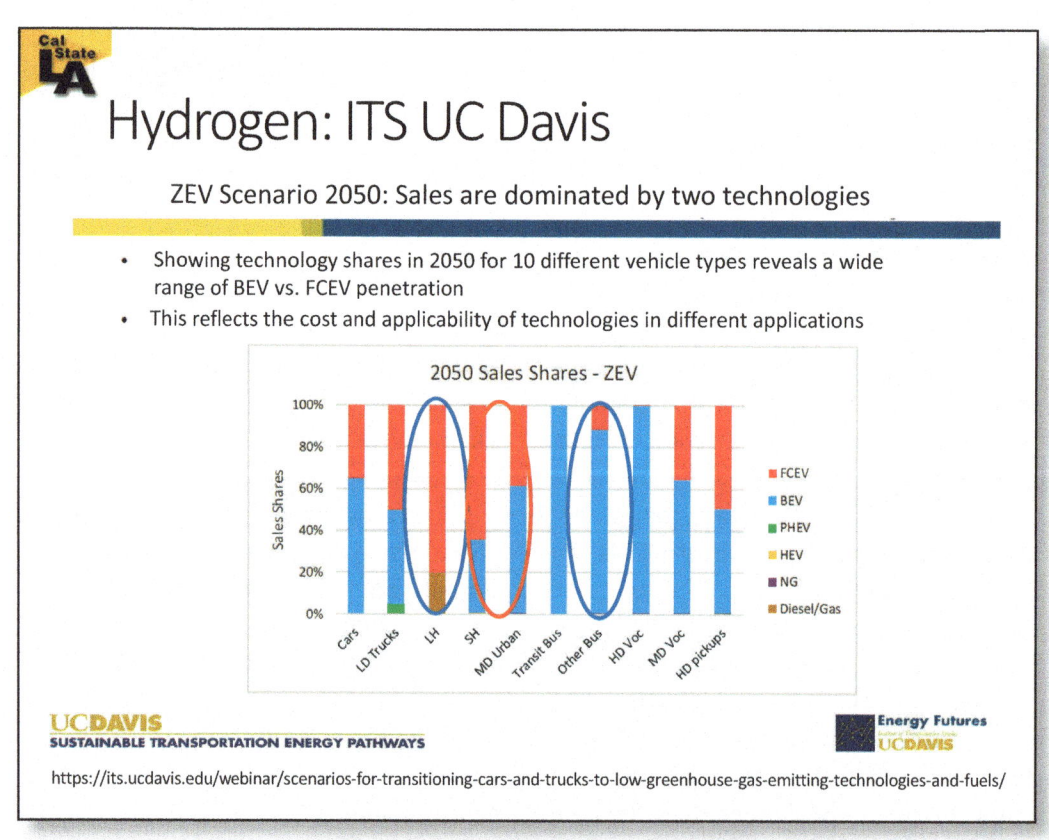

Figure 10 — **BEV:** battery EV; **PHEV:** plug-in hybrid EV; **HEV:** hydrogen EV; **NG:** natural gas; **LD:** light-duty; **HD:** heavy-duty; **MD:** medium-duty; **Voc:** Vocational use

and E-bus. I personally was impressed with the fuel cell buses running in London. Canada was also successful in demonstrating fuel cell buses during the Winter Olympic Games in Vancouver in 2010.

Over the years, there have been attempts to launch fuel cell trucks—with Vision Motors as an example that failed, being too early to the market. Nikola has been in the news with their beautiful trucks but has been plagued by some questionable business practices. The early-market introductions have been supported by newcomers like Toyota, Hyundai, Refire, Hyzon, Gaussin, and others. Volvo and Daimler formed an alliance to produce their own trucks, which was reflected in a *Forbes* article after my work in Sweden, where I had an opportunity to learn more about Volvo's aspirations. The article was titled "Volvo's Fuel Cell Truck Alliance with Daimler Is a Return to the Hydrogen Bandwagon."

Hyundai has come out with a number of projects that deploy hydrogen trucks in their XCIENT heavy-duty line. One of them is the deployment of seven trucks in Switzerland, where fuel cell trucks have perfect conditions working in the mountainous regions, as their battery counterparts would have challenging times climbing those hills over a distance. Thirty more trucks will be coming to Northern California for Glovis America, a logistics service

Eldorado bus (above), E-bus (below)

Hino truck (above), Nikola truck (below)

Figure 11

Figure 12

provider operating at the Port of Oakland. These trucks are equipped with 700 bar storage, allowing a 500-mile range.

Gaussin, a French technology company, has turned out to be a very unusual entry into the logistics and heavy-duty fuel cell transportation market. It is bold and ambitious. Just to prove their confidence, in one year they put together the first-ever hydrogen truck to compete in the 2022 Dakar Rally, in which the team was successful (**Figure 12**). Gaussin's history, aspirations, and products are so remarkable that I put together two recent *Forbes* articles, one concentrating on the Dakar Rally and the other on the company's logistics products: "Dakar Rally to See Its First Hydrogen Truck Entry by Gaussin in Partnership with Aramco" and "Gaussin's Hydrogen and Electric Transport for Cargo at Ports, Logistics Centers, and Airports."

A Colorado-based company, Vehicle Projects, was involved in a number of innovative heavy-duty hydrogen projects in the early years of hydrogen. One of them was a hydrogen train used as a switch locomotive by BNSF at its switchyard in the city of Commerce, California, demonstrated in 2011. In this application, the 127 t locomotive was powered by a hybrid fuel cell–battery system. Batteries provided low-demand and peak power, up to 1 MW, while the 240 kW Ballard PEM fuel cell continuously recharged them. Hydrogen storage was 70 kg at 350 bar. The locomotive demonstrated demanding switching and shunting for up to 10 hours per fueling.

Light rail has also been very actively pursued by several manufacturers, with many trains delivered by Alstom. Fuel cell trains provide convenient, zero-emissions transport without the burden of costly electrification, and thus have a strong advantage. requiring the hydrogen infrastructure only at a few refueling points.

Perhaps surprisingly, the industry that holds the most hydrogen fuel cell vehicles is the warehouse logistics sector. The last time I looked at the numbers, there were tens of thousands of forklifts deployed. The value here is that fuel cell forklifts are a natural fit for the industry, with most of the fuel cell advantages shining brightly: Fuel cells refuel fast and can operate 24/7 indoors with no pollution. Battery-based forklifts, on the other hand, need to recharge for hours or require special rooms for swapping batteries and recharging them, taking up valuable space and personnel.

As my students and I were imagining hydrogen tugboats in 2017, the marine industry started to think about real projects as well. Concurrently, the fuel cell industry responded by developing fuel cell packages that could be seaworthy, such as those now available from Ballard and PowerCell. There are a number of projects around the world with smaller ships, such as ferries and pusher boats, under construction or even recently completed. Havila Kystruten, a Norwegian shipping company that sails the Norwegian coast, is developing a hydrogen-powered cruise ship with a 3.2 MW fuel cell on board. This would offer a fantastic option for tourists.

As a matter of fact, Norway is very much involved in hydrogen projects through its research at SINTEF and demonstration project Haeolus, a 2 MW electrolysis hydrogen production plant powered by wind. Kawasaki Heavy Industries of Japan is designing a liquid hydrogen carrier ship, as Australia and other countries aspire to supply the world with clean fuel. This aligns well with many turbine manufacturers' efforts to convert their turbines to run on hydrogen—for example, Mitsubishi, which might develop a power plant for those carrier ships.

A port can be envisioned as a regional hydrogen hub with a multitude of functions related to it. It can include renewable energy generation; on-site production and fueling; and maritime, logistics, and ground transportation applications. Similarly, these concepts can be applied to airports and similar environments.

Already in 2019, I had discussions with the Antwerp Port Authority about their project with a tugboat powered by a diesel engine converted to run on hydrogen. Other ports around the world are also coming forward with various hydrogen deployments that we could soon see at the Port of Rotterdam in The Netherlands, the Port of Valencia in Spain, the Port of Los Angeles in the US, and elsewhere.

In an exciting project that took several years of design and manufacturing, San Francisco is to see *Sea Change,* a hydrogen ferry, operating Bay routes. Switch Maritime completed construction and started

Energy Observer's Visit to California

Figure 13

sea trials of the ferry in November 2021. The ferry has 360 kW of fuel cell power supplied by Cummins and has a capacity to hold 246 kg of hydrogen. Notable is that the first fueling took place from a hydrogen trailer that contained hydrogen produced at the Cal State LA Hydrogen Research and Fueling Facility—two states away from Washington state, where the vessel was built. As the project moved forward, fuel availability and logistics were addressed.

When the *Energy Observer* docked in Long Beach, California, in April 2021, as in **Figure 13**, it was exhilarating to be invited to welcome the crew to California and tour the magnificent catamaran yacht, which in turn is a renewable energy laboratory. The ship is on a world tour to demonstrate green technologies and to conduct research on ocean life. Thus, the crew consists of engineers and marine biologists working side by side.

In many ways, this ship reminded me of the Nilsson energy house we considered earlier. It has solar energy coming in from five different types of solar cells, and the electricity produced electrolyzes water to hydrogen, which is stored and used overnight for propulsion. Of course, this is a far more complex application and has to work in a much more demanding environment than the house. For example, Cal State LA provided 100 gal of purified water for the electrolyzer while the ship's desalination unit underwent a repair. Being a demonstration ship, it also switched its original experimental fuel cell to a trusted Toyota-sourced unit. Besides being the name of the ship, Energy Observer is a French company

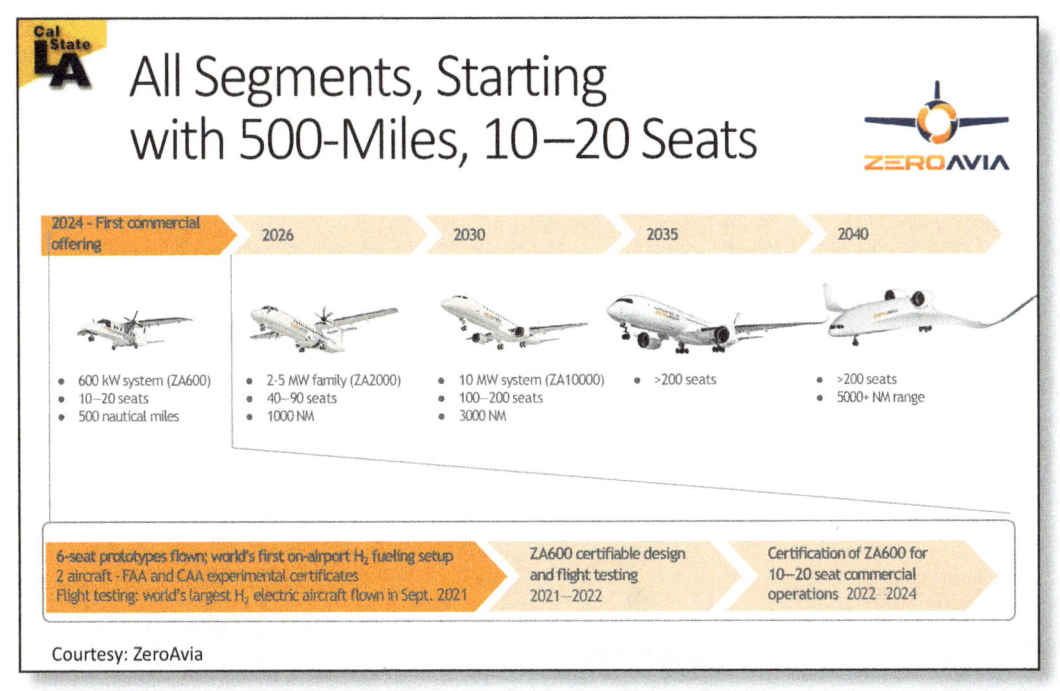

Figure 14

that focuses on fuel cell and other green technologies.

It is hard to believe, but aviation is also fair game for hydrogen deployment. As other industries turn green, aviation might be accounting for a much larger portion of emissions if it does not change its ways. Thus, ZeroAvia not only imagines but realizes fuel cell–based propulsion in the air. In recent announcements, Airbus also proposed using hydrogen planes, but its work is based on using turbines burning hydrogen. Indeed, fuel cells combined with the required electric motors might be a hefty package. To resolve this issue, ZeroAvia has partnered with Hypoint, a high-temperature hydrogen fuel cell company. Their technology relies on phosphoric acid electrolyte, which allows simplifying some systems and making the fuel cell propulsion lighter and more feasible in aviation. To the company's credit, it also developed the ground support ecosystem that includes sourcing the hydrogen and fueling the planes it develops.

In September 2020, ZeroAvia "completed the world's first hydrogen fuel cell–powered flight, which took place at the company's R&D facility in Cranfield, England, with the Piper Malibu six-seat plane." The flight was 8 minute long and climbed to 1000 ft. The company believes that hydrogen can power all sizes of aircraft with larger frames, benefiting from liquid hydrogen storage on board. The company has been gaining support and securing partnerships with airlines seeking cleaner solutions as well, as shown in **Figure 14**.

In June 2021, under new US Secretary

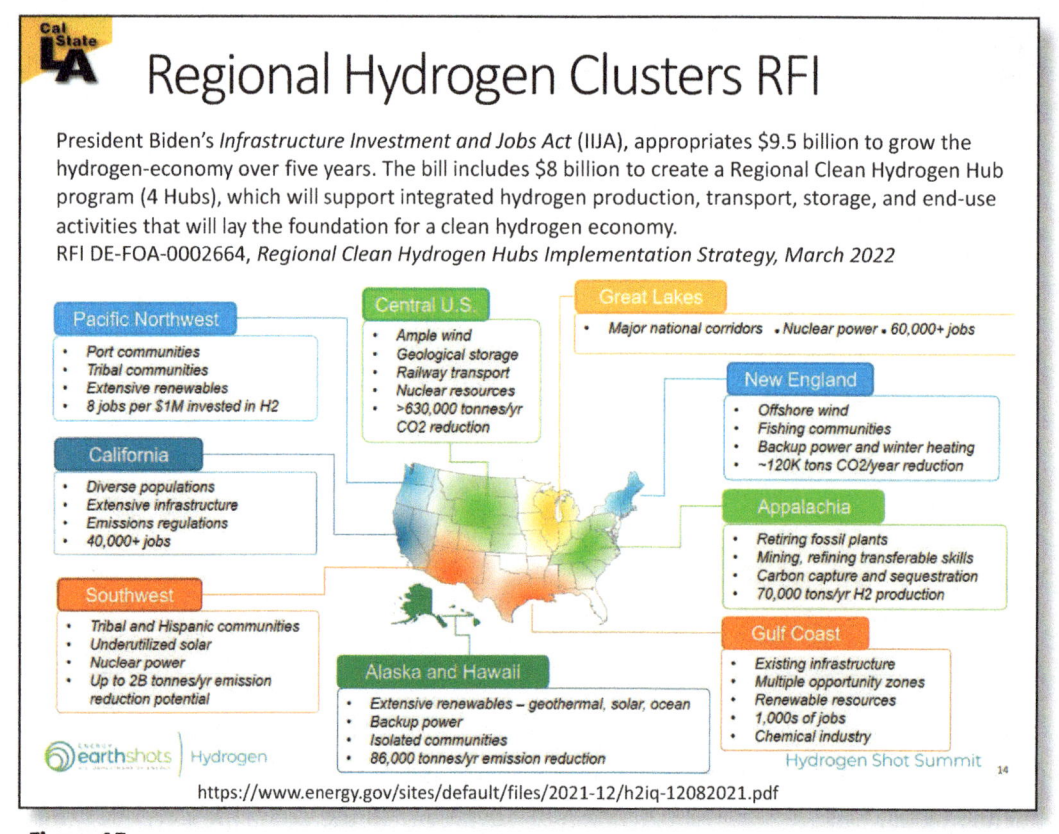

Figure 15

of Energy Jennifer Granholm, the Department of Energy (DOE) launched the Energy Earthshots Initiative to accelerate breakthroughs in more abundant, affordable, and reliable clean energy solutions within the decade. The first Energy Earthshot is the Hydrogen Shot, launched "to look beyond incremental advances and aim, instead, at the game-changing breakthroughs… to enable low-cost, clean hydrogen at scale." Under this initiative, DOE is funding projects that will address hydrogen generation with more affordable electrolyzers and other hydrogen pathways that will assure hydrogen production of 1 kg for US$1 in one decade. This, of course, will help to respond to the crucial question of the future economy: What will be the affordable nonfossil fuel?

President Biden's Infrastructure Investment and Jobs Act, signed in November 2021, will appropriate US$9.5 billion to grow the hydrogen economy over five years. The bill includes US$8 billion to create a Regional Clean Hydrogen Hub program, with four hubs in five years. A hub will support integrated hydrogen production, transport, storage, and end-use activities in a selected region, which will lay the foundation for a clean hydrogen economy. Previously, DOE had issued a Request for Information (RFI) to assess Regional Hydrogen

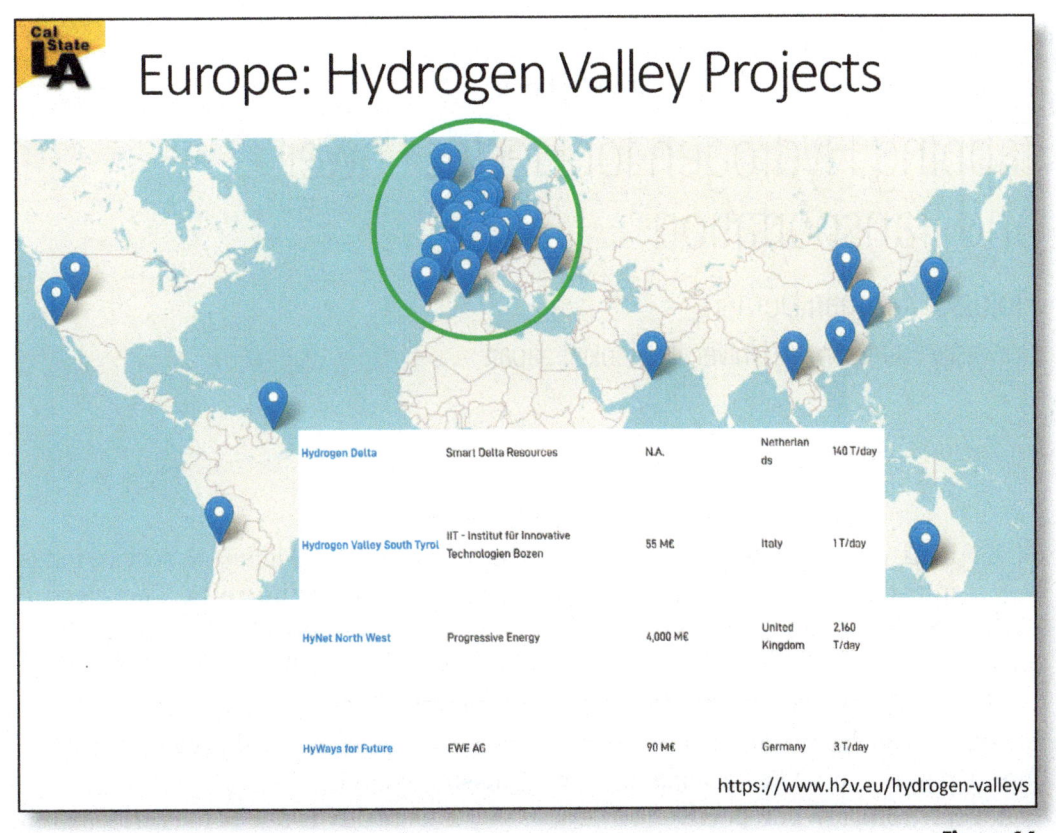

Figure 16

Clusters, with the results recently presented. **Figure 15** shows the nine interested regions. This is now being followed by a similar RFI for potential hub regions and teams. Establishing hydrogen hubs is an ongoing effort that will be exciting to watch over the next 5–10 years.

The hub efforts in the United States are slightly behind those in the European Union. It seems that Europe started later but drives much faster. Germany and Japan already have more than 100 hydrogen stations each. Under the guidance of the Fuel Cell and Hydrogen Joint Undertaking, Europe has been funding a relatively large number of hydrogen valley projects, for which descriptions can be found on the H2V website (www.h2v.eu). A simple look at the map in **Figure 16** shows that we are to see many hydrogen projects spreading around the world and creating opportunities for many countries and industries. Please extend your warm welcome to the hydrogen economy! ◊

Session 1 | First Commentary
on the Presentation by Professor David Blekhman

Tapping Hydrogen for Electric Power and Transportation

Professor Kazunari Domen
University Professor, The University of Tokyo, Japan

Thank you very much for your kind introduction. I also thank Prof. Blekhman for his very nice lecture concerning hydrogen energy utilization.

My field is a little different. I am working on solar hydrogen production. Therefore, today I would like to make a brief comment from my perspective. First, I need to discuss the forms of work energy used in human societies today. About 30% to 40% of the total work energy consumption is utilized as electric energy. The other part—more than half of all the energy used—is thermal energy. This thermal energy is consumed by the steel industry, chemical industry, service industry, and so on (**Figure 1**). The electric energy can be provided by several different resources, such as fossil, nuclear, and renewable. On the other hand, the thermal energy is provided predominantly by fossil resources.

This reality is one of the serious problems to consider in reducing CO_2 emissions. First of all, I will explain how hydrogen energy can be transformed into electric energy or transportation energy or mechanical energy and that this process itself is a clean process. On the other hand, hydrogen is derived from several different

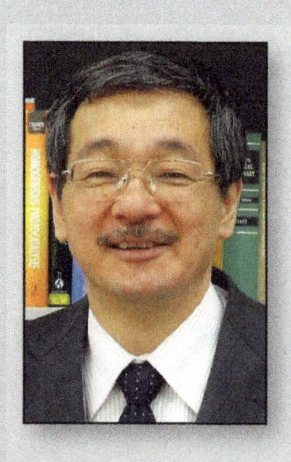

About the Speaker

- University Professor, Office of University Professors, The University of Tokyo, Japan.
- Special Contract Professor, Research Initiative for Supra-Materials, Shinshu University, Japan. (Cross Appointment)
- Researcher in photocatalysis, photocathodes, solar hydrogen production, and water splitting.
- Recipient of the Catalysis Society of Japan Awards (2007).
- Recipient of the Chemical Society of Japan Awards (2011).
- PhD in Chemistry, The University of Tokyo, Japan.

resources and is classified into gray, blue, and green energy. In the case of gray and blue energy, the hydrogen comes from fossil resources. When gray hydrogen is manufactured, carbon dioxide is actually released into the atmosphere. But in the case of blue hydrogen, produced carbon dioxide is trapped using technologies called *carbon capture and storage* and *carbon capture, utilization, and storage*.

For the time being, I believe that blue hydrogen will come to dominate the hydrogen energy market. However, as I said, blue hydrogen is derived from fossil fuels, so it cannot be considered a truly sustainable energy resource. Eventually, we have to move to green hydrogen energy, which is produced probably from water. Here, I would like to discuss just the green hydrogen production systems and their costs.

Most people probably imagine that this system should work using solar and wind power generation and an electrolyzer, since all of these technologies have already been commercialized. However, no large-scale commercial green hydrogen production systems exist in the world, except in very special areas such as Norway, where it is possible to provide very cheap hydropower. This lack of production is due to the cost of green hydrogen being much higher than that of gray hydrogen.

In Japan, we have a green hydrogen system at the Fukushima Hydrogen Energy Research Field. There, we have photovoltaics and alkaline electrolyzers. This system is one of the world's largest, the production capacity being about 1200 m^3 H$_2$/h or 100 kg H$_2$/h (**Figure 2**). However, the cost for this hydrogen is still very great. Seeing this, our government

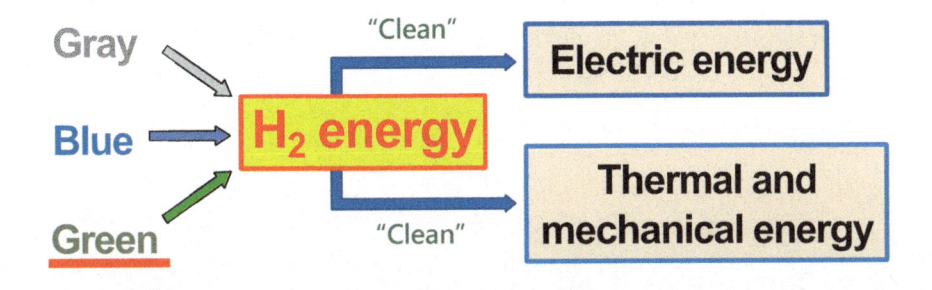

Figure 1

set a target hydrogen price. By 2050, the cost should be about ¥20/m³. Also, in the United States, the Department of Energy (DOE) set the hydrogen price at US$2/kg H_2 a few years ago. If we compare these two prices in equivalent units, while it is more costly per unit of hydrogen in Japan— about ¥0.5/molH_2 versus the US cost of about ¥0.44/molH_2 —it is almost the same.

However, as Prof. Blekhman also mentioned, the DOE last year released a new project, the Hydrogen Energy Earthshot, which targets a hydrogen cost of US$1/kg H_2 in 10 years. Nonetheless, as I said previously, the technology to produce green hydrogen is already commercialized and is more or less mature. Therefore, it will be very difficult to reduce the cost of this kind of hydrogen. I believe that we need to make some innovative breakthroughs, as otherwise it will not be possible to achieve these numbers.

Now, let me just briefly introduce our approach in Japan. We are now working with the so-called ARPChem Project, which is a 10-year project that finished last February. It was supported by the Ministry of Economy, Trade, and Industry. In this effort, we did not use any electric power but split the water directly into hydrogen and oxygen using photocatalysis. Examples of photocatalysis are shown in **Figure 3**, using various kinds of powdered materials, and are very easy to spread to a wide area.

Figure 4 shows a typical example. This is a world-class outdoor system to split water. The light-receiving area is 100 m² and consists of 1600 of the units of this size using our water-splitting powders. In addition to that, there is a gas separation unit that contains a gas separation mem-

Green H₂ Production Systems and Their Costs

- **Solar and/or wind power generation + electrolyzer**

All technologies have been commercialized
However, no large-scale commercial "green H₂" production systems

⇒ Cost of "green H₂" is much more expensive than "gray H₂"

Example: Fukushima Hydrogen Energy Research Field
Photovoltaics + alkakine electrolyzer (Asahi Chemical Co.)
1200 m³ H₂/h ≈ 100 kg H₂/h

Target H₂ price: Japan ¥30/m³ (H₂ gas) by 2030, ¥20/m³ (H₂ gas) by 2050
USA $2/kg H₂

¥ 0.5/mol H₂ (2050: Japan)   vs ¥ 0.44/mol H₂ (USA)

DOE: Hydrogen Energy Earthshot ; $1/kg H₂ in 10 years
Budget: a few billion US $ (?)

Some innovative breakthroughs are necessary

Figure 2

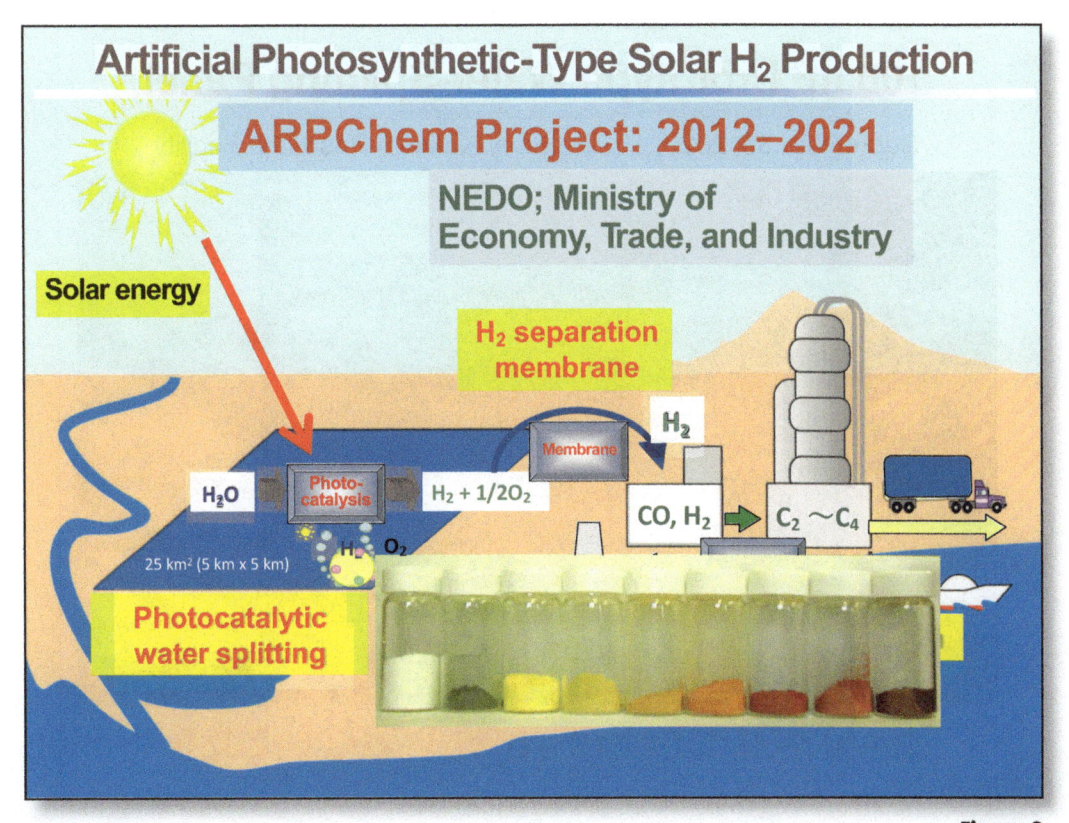

Figure 3

Figure 4

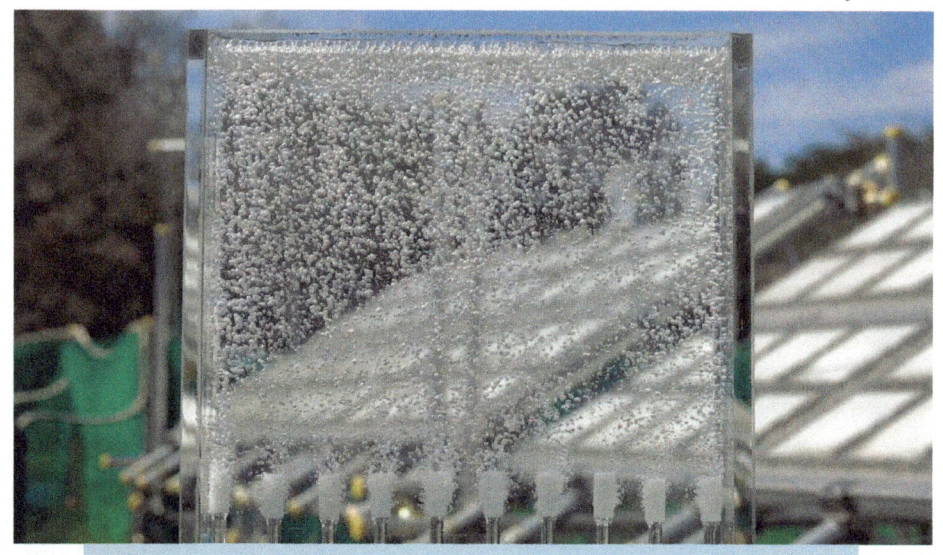

Figure 5

brane for isolating hydrogen from the mixture of hydrogen, oxygen, and water vapor.

Moving on to **Figure 5**, we see a photo showing the hydrogen and oxygen mixture evolved from this system before introducing it into the gas separation unit.

Just last March, we started our new ARPChem project. This is another 10-year endeavor. Its target is to build a practical system to produce solar hydrogen using photocatalytic systems. The target cost of hydrogen is very similar to that of the DOE's new project. My message and my comment are that researchers working with green hydrogen production should focus their efforts to produce very cost-effective, low-cost solar hydrogen, or green hydrogen, in the near future on a large scale. Thank you very much for your kind attention. ◊

References

Nishiyama, H., T. Yamada, M. Nakabayashi, Y. Maehara, M. Yamaguchi, Y. Kuromiya, Y. Nagatsuma, H. Tokudome, S. Akiyama, T. Watanabe, R. Narushima, S. Okunaka, N. Shibata, T. Takata, T. Hisatomi, and K. Domen. 2021.
"Photocatalytic Solar Hydrogen Production from Water on a 100-m^2 scale."
Nature 598: 304–307. Accessed September 20, 2022.

Session 1 | Second Commentary
on the Presentation by Professor David Blekhman

Tapping Hydrogen for Electric Power and Transportation

Dr. Dave Edlund
CEO, Element 1 Corp
Bend, Oregon, USA

Prof. Blekhman has shared a comprehensive and interesting overview chronologizing the advancing commercialization of fuel cell technology over the past few decades, especially in the transportation sector. However, as we contemplate tapping hydrogen for electric power, I think it is important to consider where that hydrogen will be sourced, the associated economics, and the carbon intensity. My commentary is focused on addressing this topic with an emphasis on land-based transportation. Accompanying this written commentary are three slides (referred to numerically in this document and presented in the paper).

Seldom, if ever, does a single product or technology win for all relevant applications. Rather, commercial solutions draw from a mix of technologies and products, and I'd venture that the same will be true as we look to electrifying transportation.

Referring to **Figure 1**, it is helpful to think of various modes of transportation in terms of energy consumed between recharging or refueling events. Energy is expressed in units of kilowatt-hours or megawatt-hours. At low kilowatt-hours, batteries are a convenient and affordable solution for electrifying transportation.

About the Speaker

- Cofounder and CEO, Element 1 Corp, USA.
- Has worked on developing fuel cells, hydrogen generation technology, and hydrogen purification membranes for 35 years.
- Board Member, e1 Marine, USA.
- PhD in Chemistry, University of Oregon, USA.

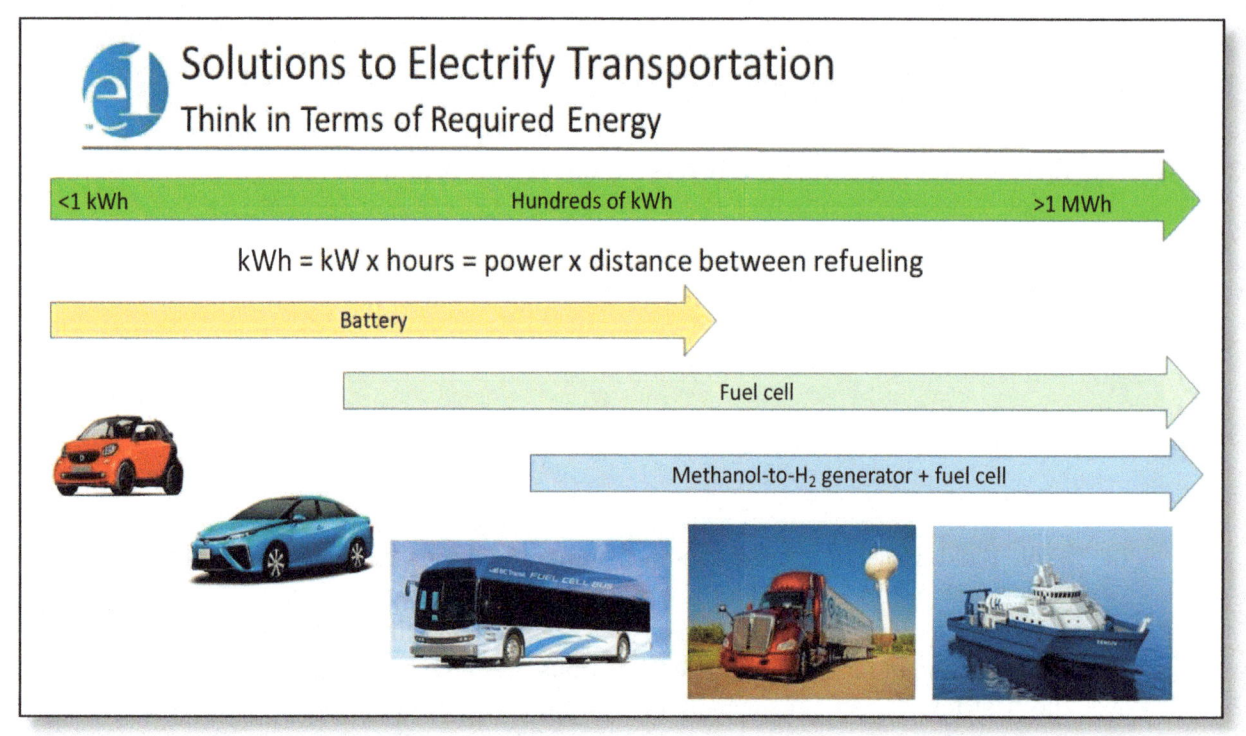

Figure 1

At a few hundred kilowatt-hours of energy, fuel cells and compressed hydrogen begin to show advantages in range. Certainly, at megawatt-hours of energy, batteries are problematic, and this space will be dominated by hydrogen fuel cells. However, storage of megawatt-hours of compressed or liquid hydrogen also presents challenges, and sound arguments can be made for conversion of methanol to hydrogen onboard the vehicle (or vessel) rather than hydrogen storage.

As Prof. Blekhman has shared, hydrogen fuel cells are gaining market share in land-based transportation as well as maritime applications. Fuel cell electric vehicles (FCEVs) have undergone global growth by segment. However, the rollout of FCEVs has been constrained by the limited number of hydrogen fueling stations. In the United States, the largest concentration of hydrogen fueling stations is in metropolitan Los Angeles (Southern California), and these stations largely dispense gray hydrogen made from natural gas or electricity from the national grid. These stations are expensive to build and may be challenging to site due to zoning restrictions and safety setbacks.

In contrast to FCEVs, battery electric vehicles (BEVs) are popular, with all major automobile manufacturers offering at least one model. In the United States, the sale of BEVs has outpaced the sale of FCEVs by more than 40 to 1, with a forecast of nearly 5 million BEVs to be sold in the

Why Methanol as a Precursor to Hydrogen?
More Hydrogen Than in Liquid H$_2$

Methanol is the superior hydrogen carrier

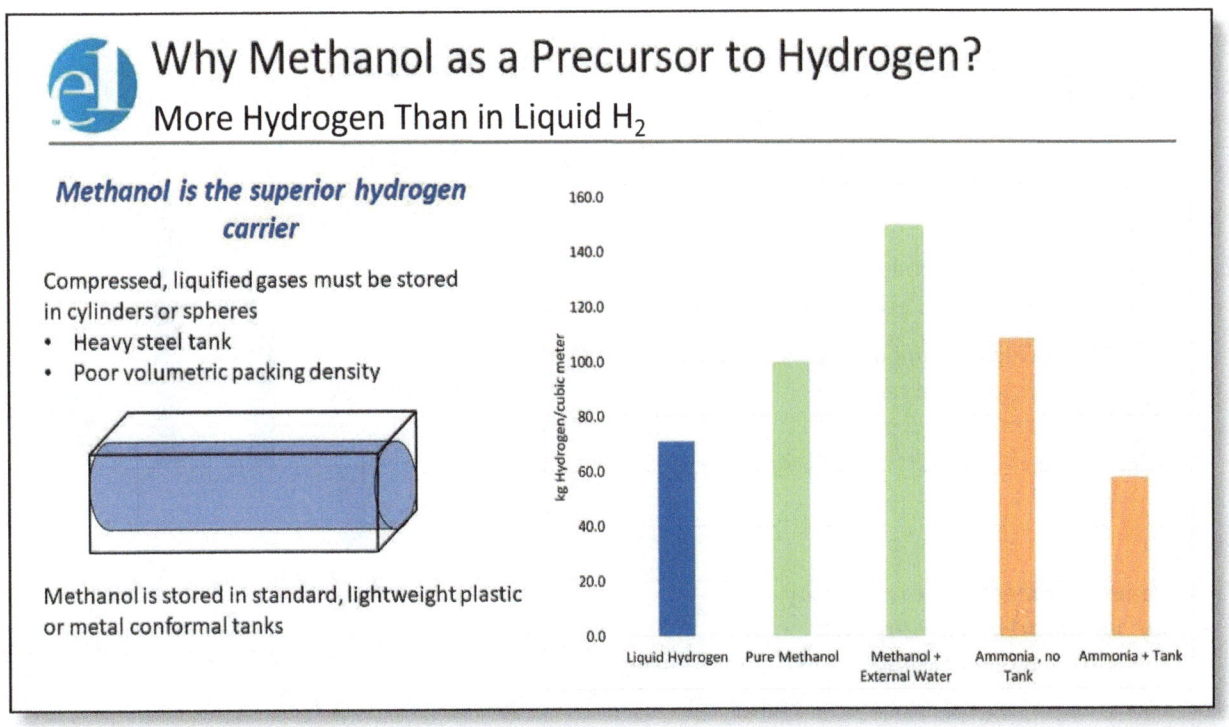

Compressed, liquified gases must be stored in cylinders or spheres
- Heavy steel tank
- Poor volumetric packing density

Methanol is stored in standard, lightweight plastic or metal conformal tanks

Figure 2

United States in 2030. Given the popularity of BEVs, it makes sense to consider the question of BEV charging infrastructure. Although there are many times more public BEV charging stations than there are hydrogen fueling stations, the infrastructure build-out for charging stations is still an enormous challenge that is hindered by national electrical grids that have little room to expand capacity.

One way to address the need to build out infrastructure to support both compressed hydrogen fueling of FCEVs as well as BEV charging is to use hydrogen made on site from methanol mixed with water. Chemical conversion of methanol plus water to low-purity hydrogen (syngas), followed by subsequent purification of hydrogen, provides an on-site source of high-purity hydrogen for compression and storage, or for conversion to electricity in a fuel cell. As shown in **Figure 2**, pure methanol contains more hydrogen in a given volume than the same volume of liquid hydrogen, making methanol an attractive hydrogen carrier. Even better, adding supplemental water to the methanol increases the hydrogen yield by 50%, and the corresponding volumetric hydrogen content is twice that of liquid hydrogen. Ammonia is occasionally suggested as a hydrogen carrier, but when the heavy cylindrical tank for storing liquid ammonia is accounted for (ammonia is a compressed liquefied gas), it is less attractive as a hydrogen carrier compared

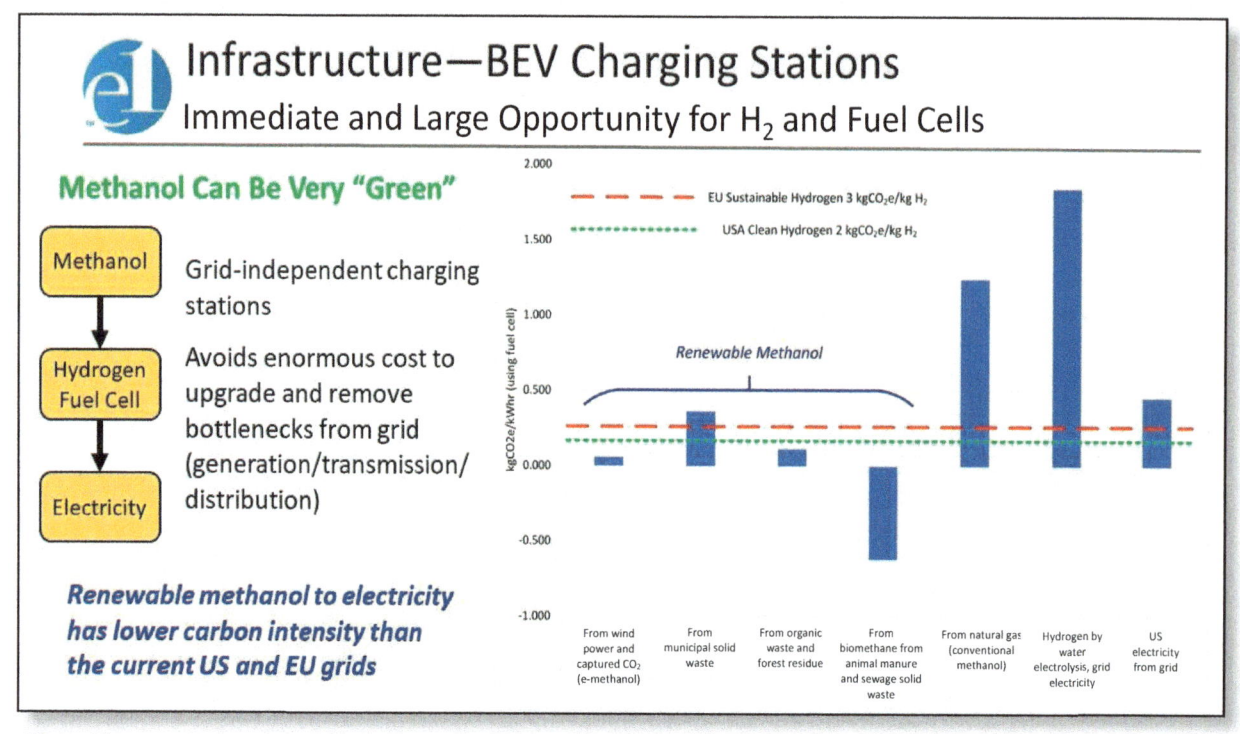

Figure 3

to methanol. Plus, ammonia is disadvantaged by several unique safety concerns such that placing large quantities of liquefied ammonia in dense urban locations will be challenging (at best).

Equally important, the cost of making high-purity hydrogen from a methanol/water mix at the point of use (such as a hydrogen fueling station or BEV charging station) is about US$3.45/kg H_2, assuming methanol is delivered at US$500/t. This is attractive and compares favorably to the pump price for gasoline in the United States.

This commentary would not be complete without addressing the carbon intensity of various pathways to hydrogen. The graph in **Figure 3** shows this in terms of $kgCO_2e/kWh$ (assuming hydrogen is converted to electricity in a fuel cell). The highest carbon intensity results when grid electricity (United States grid, although the European grid is only slightly greener) is used to split water into hydrogen and oxygen. Natural gas to hydrogen also yields a high carbon intensity.

However, several different pathways to renewable methanol yield the greenest hydrogen, in some cases with a negative carbon intensity due to avoided methane emissions. Renewable methanol results from using biogenic feedstocks (biomass, municipal solid waste, and animal and human manure) or carbon dioxide captured directly from air or from industrial waste streams (e-methanol) to make methanol. Presently, more than 35 plants are in commercial operation globally making renewable methanol, and new plants are being announced every couple of months. The maritime sector is largely driving the market for renewable methanol, with Maersk purchasing renewable methanol under multiyear contracts from all available suppliers.

It is worthwhile to note that regardless of the source of hydrogen, grid-independent charging stations can be built out at lower cost and on a faster time scale compared to any concepts for rebuilding or expanding national electricity grids, which necessarily comprise generation plants, long-distance transmission, and distribution (substation to plug).

Thank you. ◊

Session 1 | Presentation Two

Negative Emissions Technologies in U.S. Decarbonization Pathways

Dr. Eric Larson

Andlinger Center for Energy and the Environment,
Princeton University, New Jersey, USA

I would like to speak with you today about negative emissions technologies, and in particular, the role that these might play in the decarbonization of the United States economy.

I will begin by explaining why negative emissions are needed and then describe different negative emissions technologies (or NETs). Finally, I will discuss possible roles of NETs in technological pathways for the United States to reach net-zero emissions by 2050.

I will start by reviewing the science that we understand about the relationship between global warming and the emissions of greenhouse gases, especially CO_2. In **Figure 1** is a graph from the Intergovernmental Panel on Climate Change's *Special Report on Global Warming of 1.5 °C*. I will not go into all the details here.

Basically, what this graph tells us is that the warming that we can anticipate for the world is a function directly related to the cumulative emissions of carbon dioxide that we have put into the atmosphere since the preindustrial period, starting with the mid-1800s. With this understanding of the relationship, we can estimate the remaining amount of carbon budget that we can emit before we hit

About the Speaker

- Senior Research Engineer, Andlinger Center for Energy and the Environment, Princeton University, USA.
- Senior Scientist, Climate Central, USA.
- Research that intersects engineering, environmental science, economics, and public policy on sustainable and engineering-based solutions to energy-related problems.
- PhD in Mechanical Engineering, University of Minnesota, USA.

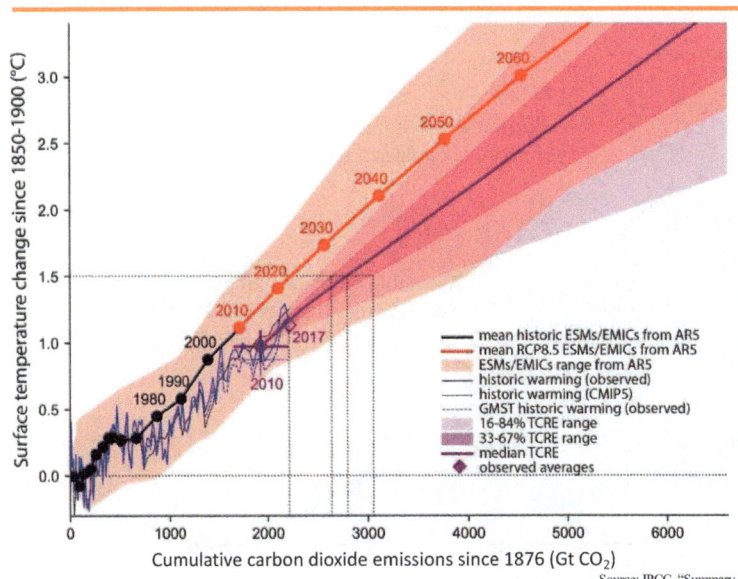

Cumulative CO₂ Emissions Determine Warming
Remaining Emissions "Budget" for 1.5 – 2°C Is Shrinking

Remaining global **carbon budget** (as of December 2018) for 50% probability of

$\leq$ 2 °C warming = ~1500 Gt CO_2

$\leq$ 1.5 °C warming = ~580 Gt CO_2

Through 2021, ~150 Gt CO_2 of these budgets has been spent

At current annual emissions rate, global budgets will be fully spent in ~10 years for 1.5 °C, and ~35 years for 2 °C

Source: IPCC, "Summary for Policymakers," in *Special Report on Global Warming of 1.5°C*, 2018

Figure 1

certain thresholds of temperature increase. As of December 2018, when this study came out, there was a 50% probability that we could stay below 2 °C warming globally if we emit no more than 1500 Gt of CO_2 cumulatively from that point forward. To stay within a 1.5 °C carbon budget, it would be, of course, much lower—closer to 600 Gt.

Since that estimate was made, we have, as a world, already emitted an additional 150 Gt. We have spent some of our budget already, which means you can estimate that if we want to stay within 1.5 °C of warming, we have about 10 years left of emissions at the current global emissions rate before we hit that threshold, and we have a bit more time if we are satisfied with staying below 2 °C.

With that understanding, we can look at a graph like that in **Figure 2**. This is based on various models and shows the trajectory of global CO_2 emissions in order to stay below the 2 °C threshold. This was done a few years back, and scientists at the time started their modeled emissions trajectory with the year 2005. Actually, as we know now, emissions from 2005 to 2015 continued to increase beyond the modeled level.

The pathway is not precise here, but it is reflective of the kind of change that the world needs to see in order to stay below 2 °C of warming: the world would need to reach zero emissions by about 2070, in other words, the world has used up its carbon emissions budget by that date.

This budget can essentially be extended

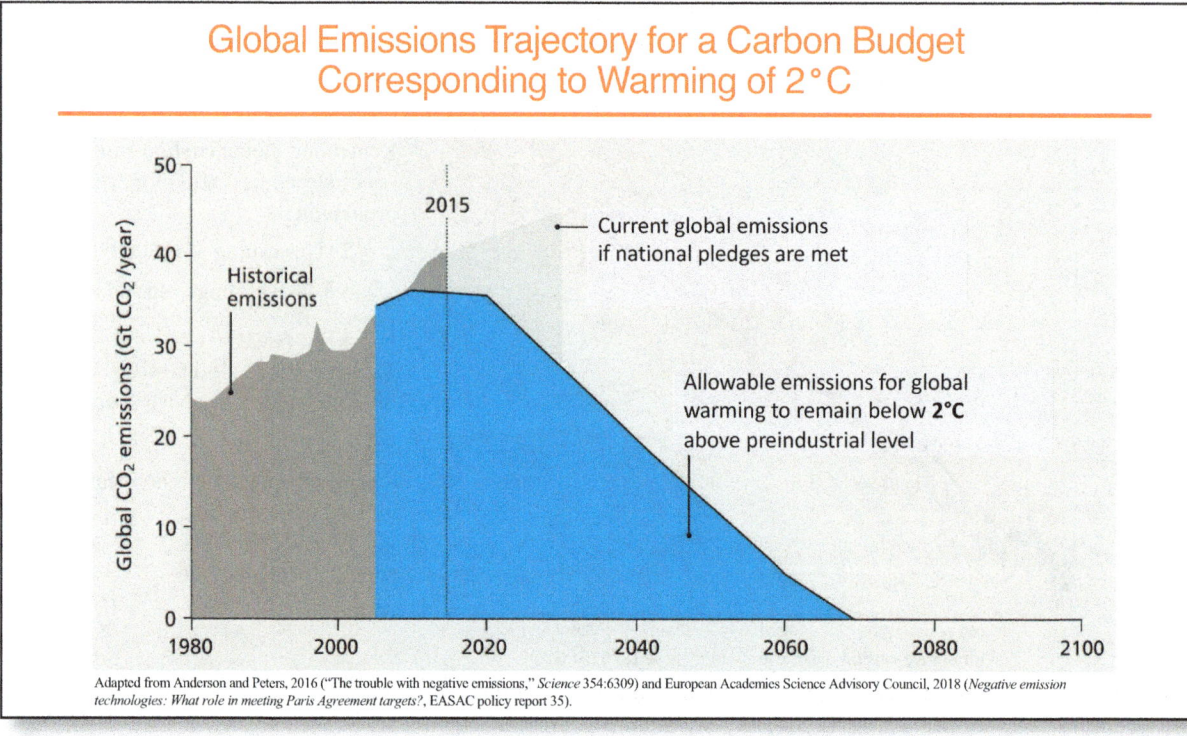

Figure 2

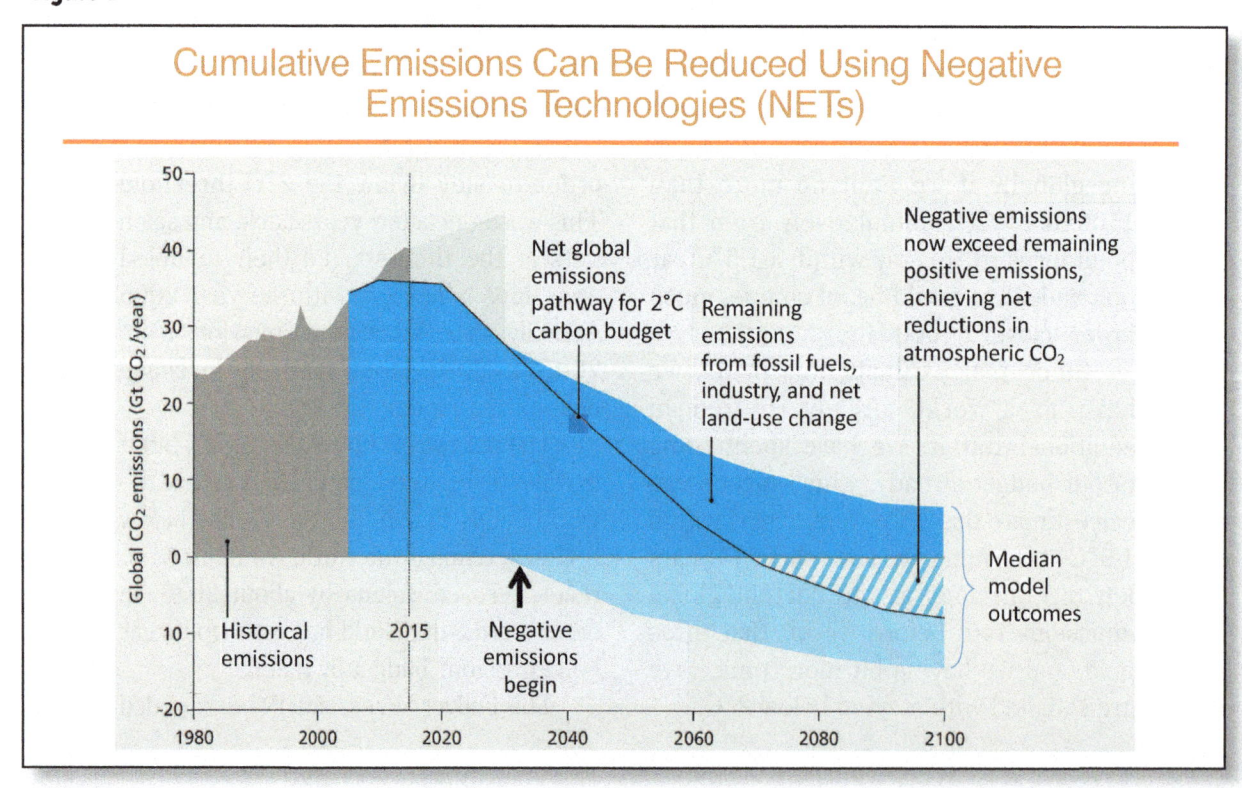

Global Emissions Trajectory for a Carbon Budget Corresponding to Warming of 2°C

Adapted from Anderson and Peters, 2016 ("The trouble with negative emissions," *Science* 354:6309) and European Academies Science Advisory Council, 2018 (*Negative emission technologies: What role in meeting Paris Agreement targets?*, EASAC policy report 35).

Figure 3

Cumulative Emissions Can Be Reduced Using Negative Emissions Technologies (NETs)

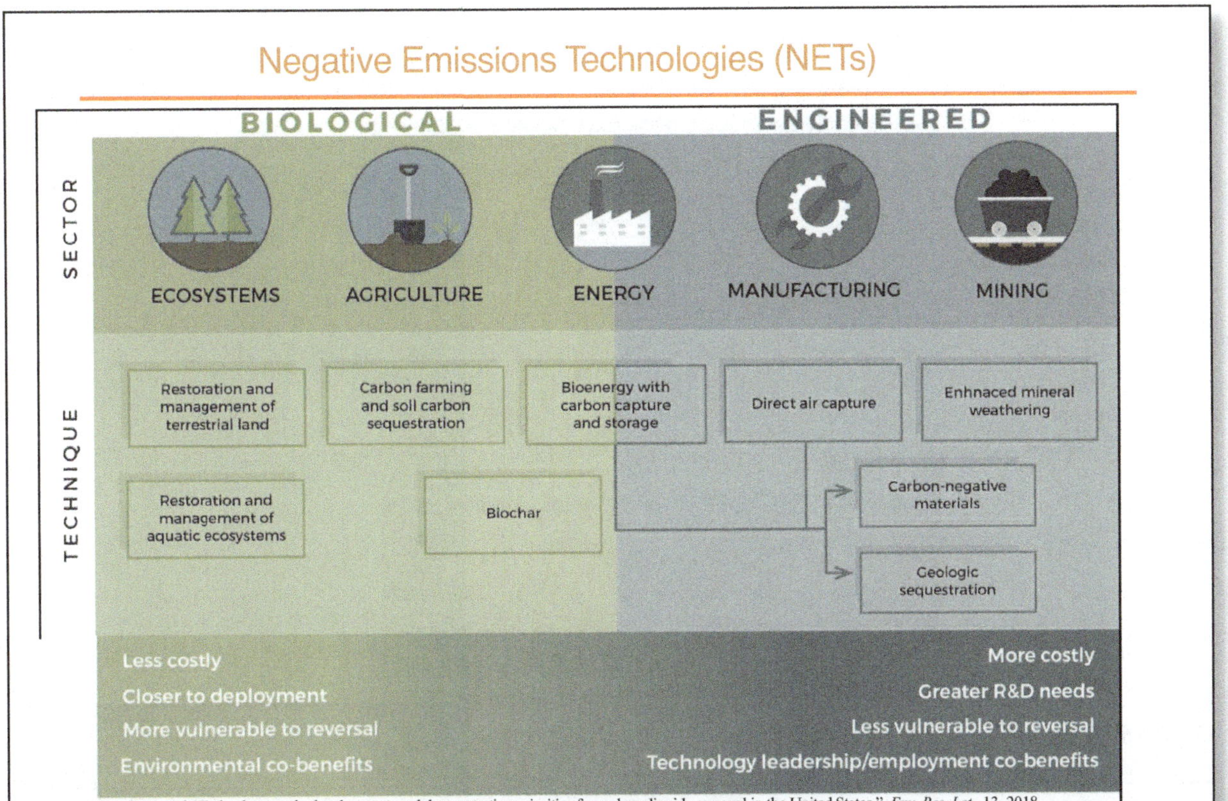

Negative Emissions Technologies (NETs)

Figure 4

Sanchez, *et al*, "Federal research, development, and demonstration priorities for carbon dioxide removal in the United States," *Env. Res. Let.*, 13, 2018.

if we allow for the possibility of negative emissions. In this case, we are on the pathway shown in **Figure 3**. Again, these are modeling results, and this pathway results in higher emissions than following the budget to 2 °C, but we compensate for that by negative emissions, beginning as early as 2030 and growing considerably beyond that. This then allows us to stay on a net-zero emissions trajectory for 2 °C.

Essentially, negative emissions allow us to increase our budget of positive emissions and to still stay below our temperature targets. There are a variety of negative emissions technologies, and we understand many of these quite well, such as those in **Figure 4**. For example, through restoration and management of terrestrial and aquatic ecosystems, we are able to absorb CO_2 out of the atmosphere. We can do the same by changing agricultural practices—so-called carbon farming. We can increase the carbon content in soils, which takes CO_2 out of the atmosphere. These are largely biological measures (left side of **Figure 4**). As we move to the right in **Figure 4**, we move toward more engineered measures, starting with bioenergy with CO_2 capture and storage. This is based on plant matter that has absorbed CO_2 from the atmosphere as it has grown. The plant matter is then

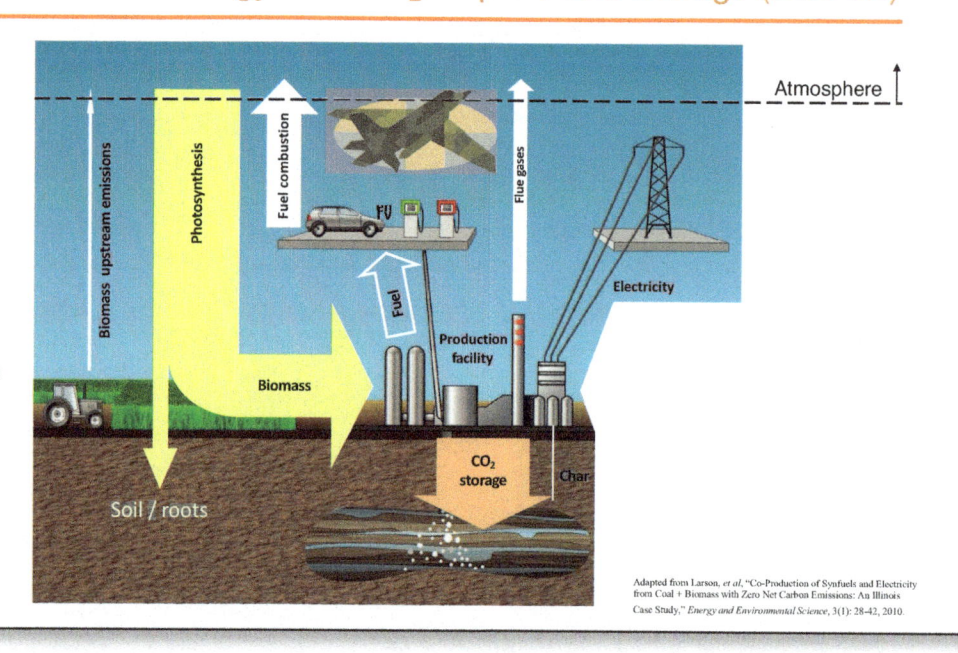

Figure 5

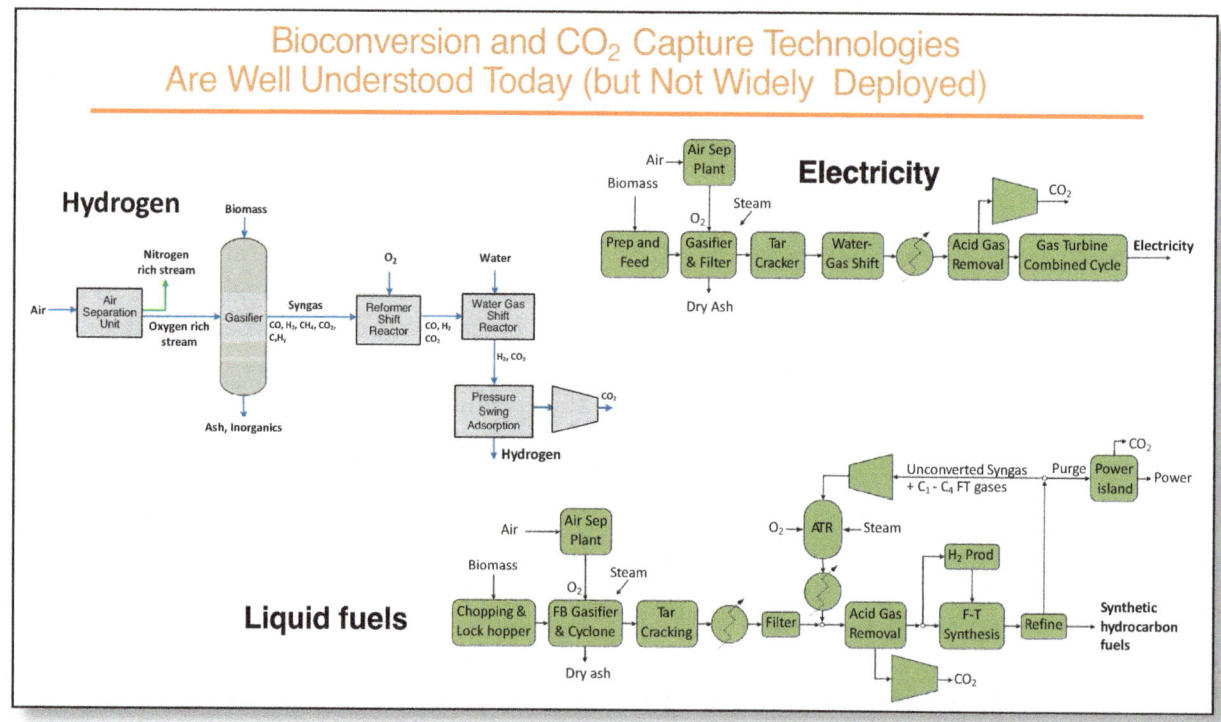

Figure 6

converted into a convenient form of energy, for example, electricity, and the by-product CO_2 of the conversion process is captured and stored underground.

More fully engineered negative emissions systems include direct air capture (DAC), where we are taking the CO_2 directly out of the air by a chemical process and then storing the CO_2 below ground. There is also enhanced mineral weathering, which is basically creating carbonate rocks using CO_2 and natural rocks that combine to make carbonate rocks and thereby store CO_2.

You can see by the notations on the bottom of **Figure 4** that biological processes tend to be less costly per ton of CO_2 that is removed. They are closer to deployment in part because they are less costly and because we know how to do these quite well. On the other hand, they are more vulnerable to reversal—that is, soil carbon can be rereleased to the atmosphere if the methods are not properly managed. That can happen quite quickly and easily. On the other hand, there are environmental cobenefits with carbon and soil that often increase the productivity of the soil, which is a positive result of that system.

Then, as we move toward the more engineered systems, we see that they are generally more costly and often need a bit more research and development and certainly more complicated deployment and demonstration of commercial capability. On a positive note, they are less vulnerable to reversal. There is also the potential for cobenefits here as well, in the form of technology leadership for countries or companies that are at the forefront in developing these potentially new employment opportunities.

Among the various negative emissions technologies, the two that are generally considered to be the most prospective in terms of the size of the role that they can play in net-negative emissions overall are bioenergy with carbon capture and storage (BECCS) and DAC with CO_2 storage.

In **Figure 5**, you can see the carbon flows for a BECCS system. The widths of the arrows in this picture are roughly equivalent to the magnitude of the carbon flows. This illustrates the fact that there are emissions at various points along this process, from tractors that might be used in the cultivation and harvesting of biomass, from unavoidable emissions at the conversion plant, and if a hydrocarbon fuel is being made, some carbon will return to the atmosphere when the fuel is used. However, a large amount of the by-product CO_2 in the conversion process is captured and put underground for storage. This carbon had been removed from the atmosphere via photosynthesis as the biomass grew. If you look at the net balance across all of the arrows, you will see there is a net flow of carbon from the atmosphere to the subsurface on an annual basis.

There are a number of different technologies for the biomass conversion process (**Figure 6**), and these are rather well understood. My group has analyzed many of these. There are other researchers around the world who also have been looking at these technologies. The challenge has been primarily over their cost, in that most of these processes are not econom-

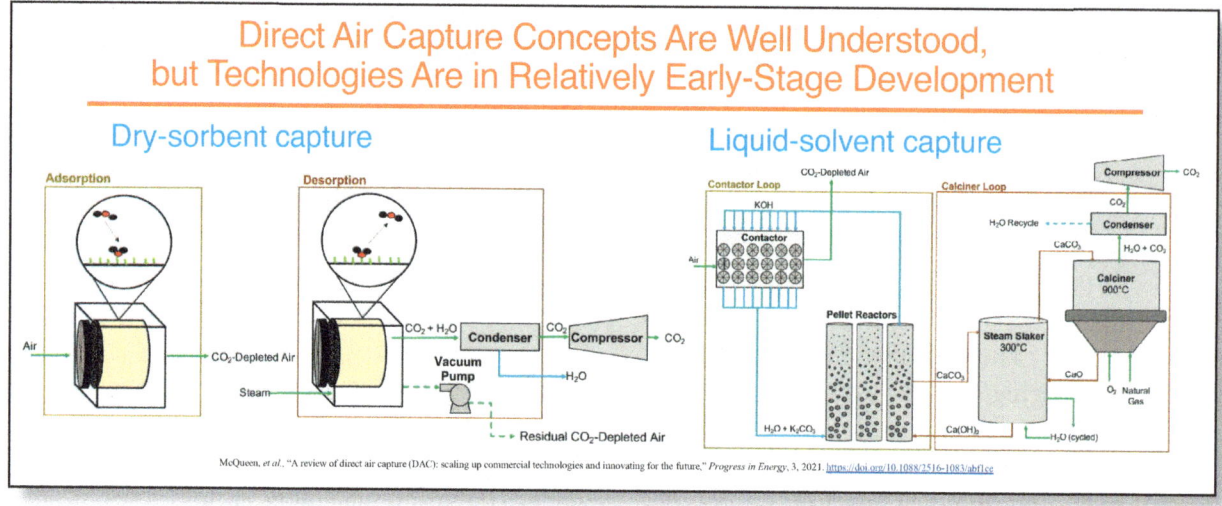

Figure 7

ical under today's conditions. Therefore, although we have a quite good understanding of how these technologies work from an engineering perspective, they are not widely deployed commercially today.

Now, let me say a few things about direct air capture. DAC concepts are also well understood, but the technologies themselves are at a relatively early stage of development. Two of the leading concepts are shown in **Figure 7**. One involves basically passing air over a dry sorbent that then selectively pulls the CO_2 molecules out of the air. The sorbent is then regenerated through some means, typically by heat addition to drive off the CO_2. That CO_2 is collected and compressed for transportation through a pipeline to an underground storage site. The scheme on the left in **Figure 7** uses such a dry sorbent. One company has now built a 4000 t CO_2/year capture facility in Iceland, starting up in the latter part of last year. That is a relatively small facility by comparison to the levels of CO_2 capture that we want in order to address the 2 °C or even the 1.5 °C challenge, but it is a start.

The other concept (right side of **Figure 7**) is centered around a liquid solvent—potassium hydroxide—that captures the CO_2 and then goes through a process to separate the CO_2 from the solvent so that the solvent can be recycled and used again. The captured CO_2 is then compressed and stored. A different company is developing this concept and has plans to have a 1 million tCO_2/year facility starting up in 2024. This begins to get to the commercial scale that will be needed in the longer term.

With both BECCS and DAC plus storage, there is a requirement for underground storage resources. Fortunately, around the world, there are many geological formations that have the capacity to store CO_2. However, the distribution of these resources varies from country to country. The United States is particularly well endowed with CO_2 storage geology. We understand this geology better than

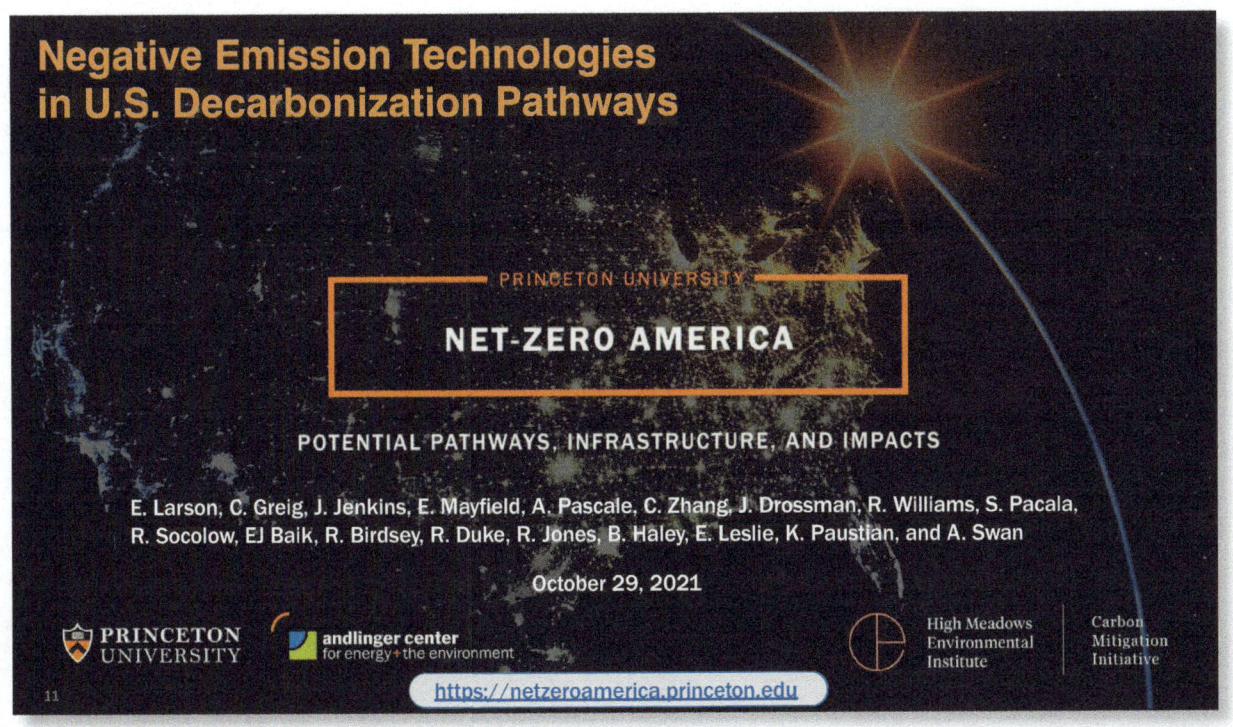

Figure 8

the geologies in many other countries because there has been work done for several decades in the US to try to understand these resources. On the order of 40 million tCO_2/year is currently being captured and stored across the world, not just in the US, via a number of demonstration projects. This points to the fact that we understand the CO_2 storage process rather well. The challenge is characterizing the subsurface sufficiently, so that one can have confidence that the storage is secure. Again, we know how to do that.

This brings me to the last part of my talk, which is the potential role that negative emissions technologies might play in the United States if the US is to achieve its government-announced goal of net-zero emissions by 2050. For this, I want to draw on a study that I co-led, which we published last year. It is available at the website shown in **Figure 8**. What we basically did was to try to paint a picture in as much detail as possible of what the US energy economy would look like if net-zero emissions were achieved by 2050.

We started with the knowledge that today's US net emissions are about 6 Gt CO_2/year. We drew a straight line for the net emissions to reach zero by 2050 (**Figure 9**). The analysis we did takes account that there is a land sink today, with trees growing and soils absorbing carbon, and that we would enhance that land sink through various measures. There were experts on our team who helped us to understand that. There are also non-CO_2 emissions that have to be

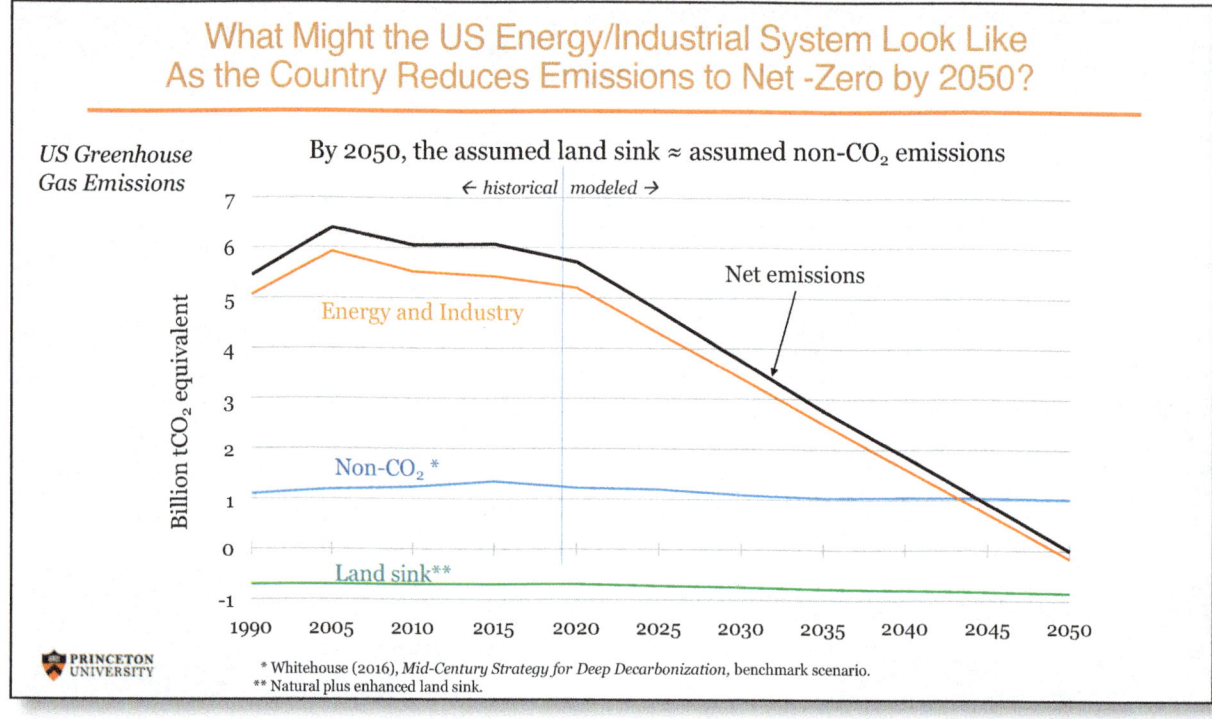

Figure 9

What Might the US Energy/Industrial System Look Like As the Country Reduces Emissions to Net-Zero by 2050?

considered, like methane and nitrous oxide that come from agricultural production. These tend to be more difficult to completely eliminate. Therefore, when you have non-CO₂ emissions and a land sink, the difference between those is what the energy system will need to provide.

In our study, we modeled the energy and industrial system and ended up with basically slightly negative emissions for those sectors by 2050 in order to meet the net-zero economywide target. We did a variety of other modeling that I will not go into detail on, but **Figure 10** shows the results of our study. This shows the primary energy supply in 2050 under different pathways to net-zero emissions. The left bar shows our current 2020 mix of energy sources of over 80% fossil fuels.

By 2050, our reference scenario (second-from-the-left bar), without any new policy measures, looks quite similar. Then there are five scenarios that all meet the target of net-zero emissions by 2050. They all deliver the same energy services—that is, the vehicle miles traveled and the square meters of building space heated and cooled and so on. However, they do this with different mixes of energy-demand and energy-supply technologies.

As an example, what we call the E+ scenario is a high-electrification scenario where buildings and vehicles are electrified very aggressively. The E− involves less aggressive electrification. As you can see, electrification gives you some efficiency benefit. In fact, an electric vehicle, for example, has maybe three times the efficiency

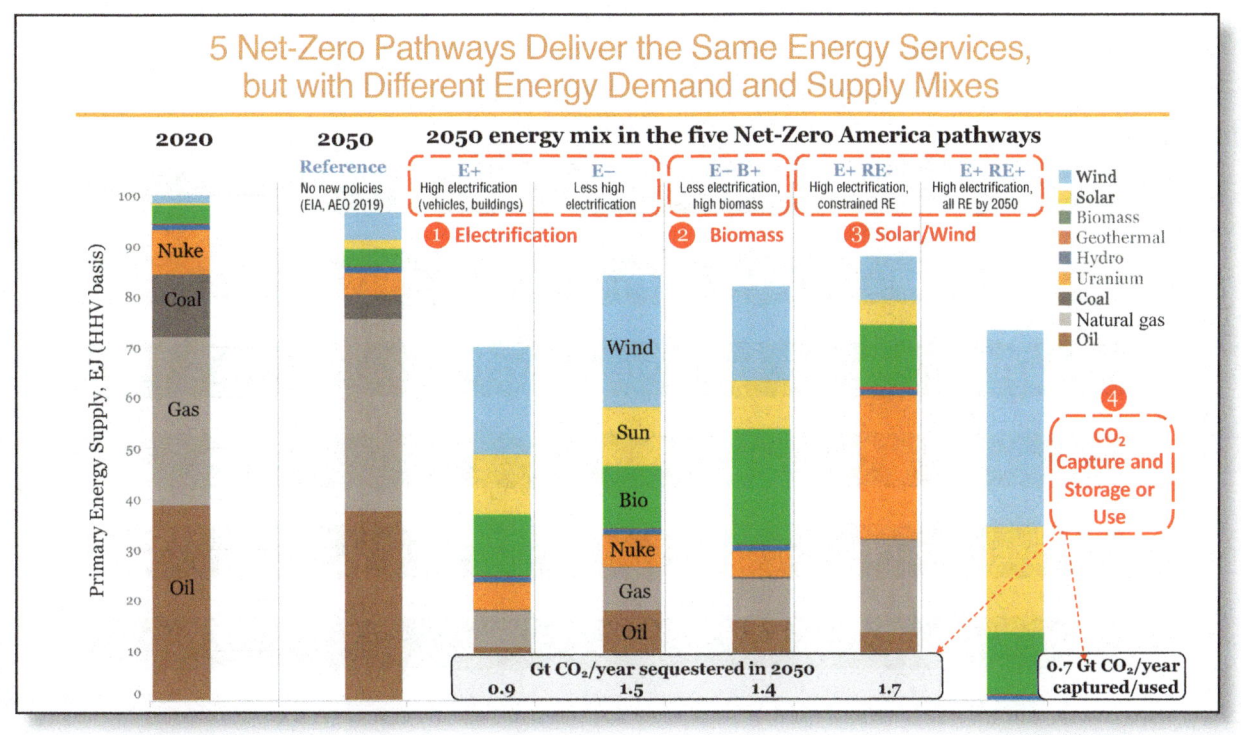

5 Net-Zero Pathways Deliver the Same Energy Services, but with Different Energy Demand and Supply Mixes

Figure 10

RE: renewable energy

of a comparable internal combustion engine vehicle, so we have less of an energy requirement overall in the E+ versus E−.

I want to point out the green bars in **Figure 10** represent biomass. In all of our scenarios, biomass is a very important contributor by 2050. Most of the biomass is used with CO_2 capture and storage as well. In four of the five scenarios (the first two and the last two of the five net-zero pathways), we limited the amount of biomass that could be used in the energy system to that which could be delivered without changing land use from today. Taking land for bioenergy that might otherwise be used for agriculture has its potential problems. Therefore, we wanted to minimize that issue. You can see that in these four scenarios, bio-energy is at about the same level.

In the fifth scenario (the middle pathway of the five net-zero bars), we relaxed the land use constraint for biomass and said, "Let us allow more biomass, including some land use change," and you can see that much more biomass is adopted there. In all five of our pathways, biomass is a very valuable energy resource when coupled with CO_2 capture and storage.

In the last two scenarios (right two bars), we changed the level of wind and solar generation. In the fourth bar, we limited the amount of wind and solar capacity that could be added annually to about 40% more than the maximum single-year addition achieved in the recent past. In the last pathway (the bar on the right), we did not place any constraint on

Unprecedented Rates of Physical Change Across Six Essential Pillars of Decarbonization

1. End-use energy efficiency and electrification
2. Clean electricity: wind and solar, transmission, firm
3. Clean fuels: bioenergy, hydrogen, and synthesized fuels
4. CO_2 capture and utilization or storage
5. Reduced non-CO_2 emissions
6. Enhanced land sinks

Figure 11

wind or solar additions, and we required the energy system to be completely fossil-fuel free by 2050.

You can see that, in the first four pathways in **Figure 10**, we still have fossil fuel use in 2050. Part of the reason we are able to continue using fossil fuels there is because we have CO_2 capture and storage involved and, in fact, in these first four scenarios, we have between 1 and 2 billion tons per year of CO_2 capture and storage. In the fifth scenario, we did not allow carbon storage, but there is still capture of CO_2, with the carbon being recycled back into fuels that area needed in the energy system.

All five of our pathways rely on six decarbonization pillars (**Figure 11**) deployed at unprecedented rates. Unprecedented means we have not seen such rates of change historically in the US; it does not necessarily mean they are impossible rates of change.

Our study delved into each of these important pillars. Here, I will show you a bit of a snapshot of some results for the BECCS and DAC technologies, in other words, the engineered negative-emissions technologies included in our model.

We did detailed mapping and prospective siting of bioenergy facilities. **Figure 12** displays our map for 2050. We did this mapping in five-year time steps. I am just showing you the 2050 map. The green point sources represent bioenergy conversion with CCS. You can see that the sites are widespread, particularly around the Midwest, but also in the Southeast as

For the Five Net-Zero Pathways, Annual Capture at BECCS Facilities Ranges from 0.4–1.5 Gt CO_2 in 2050

- 929 million t CO_2/year stored in total in 2050 (1.3x current volume of U.S. oil production)
- 106,000 km CO_2 pipelines in 2050 (equivalent to ~22% of today's natural gas transmission pipeline system)

CO_2 point-source type
- CO_2 point sources
- BECCS - power and fuels
- Cement with CCS
- Natural gas power CCS oxyfuel

CO_2 captured (MMTPA)
- 0.0006449
- 7.9144
- 15.8282
- 23.7419

Trunk lines (capacity in MMTPA)
- < 100
- 100 - 200
- > 200

CO_2 Storage Basins
2020 CO_2 Pipelines

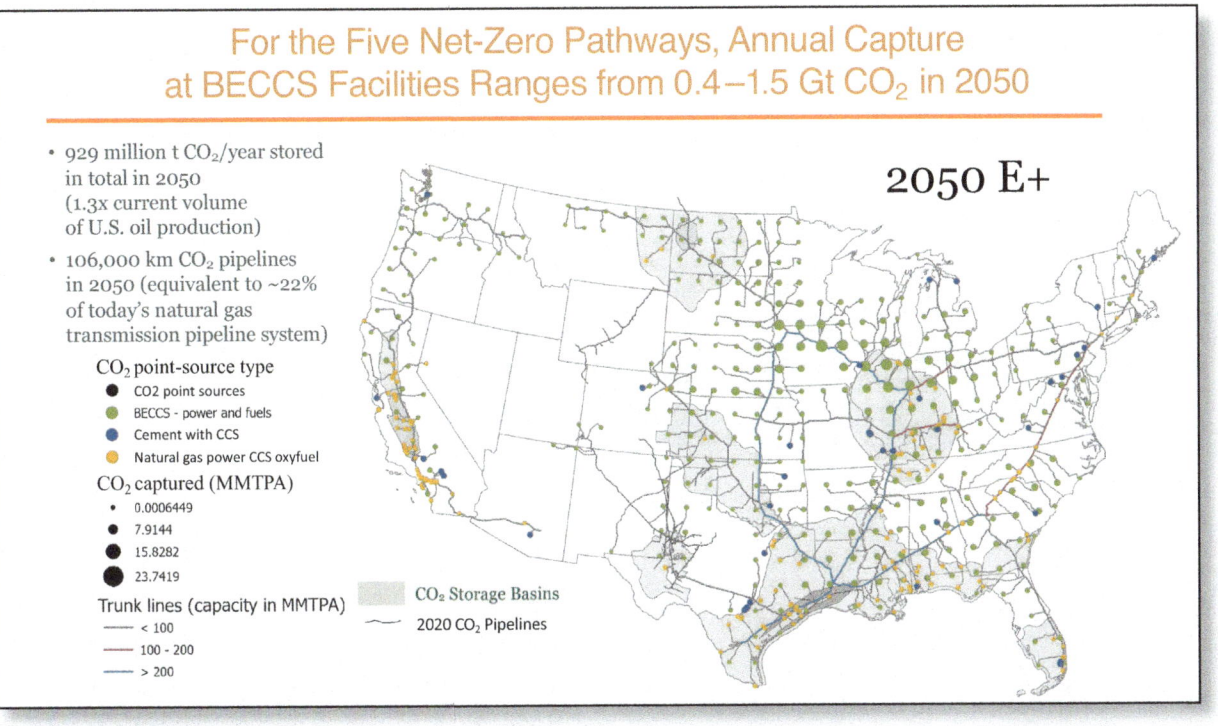

2050 E+

Figure 12

well as along the western part of the US.

We designed a pipeline network for CO_2 collection and transportation to underground storage locations. The gray-shaded regions in **Figure 12** represent the most prospective regions for CO_2 storage in the country. The volume of CO_2 capture and storage that is going on in the E+ scenario, which is not our most aggressive one, is comparable to total current US oil production, which gives you an indication that CO_2 capture, transport, and storage is a very significant new industry in our net-zero pathways.

In the upper panel of **Figure 13**, we see all the sources of CO_2 capture in 2050 in each of our five net-zero scenarios. The lower panel in **Figure 13** shows where captured CO_2 (from all sources) goes. The gray in the lower panel represents CO_2 stored. Most of the CO_2 that is captured in the first four scenarios is stored underground. You can see that bioenergy with CCS (BECCS) plays a big role in all five of these scenarios. DAC really comes in only in the E– scenario. If you recall, that is the case where we did not electrify vehicles and buildings as aggressively as we did in the E+ pathway. That leads to additional fossil fuel use in 2050, that then the emissions from those need to be offset. Since we have consumed the full biomass potential, here we need to adopt DAC, which in our modeling is a more expensive CO_2 removal option than BECCS, and that is why it comes in later and mainly in that one scenario.

Thus, both the BECCS and DAC tech-

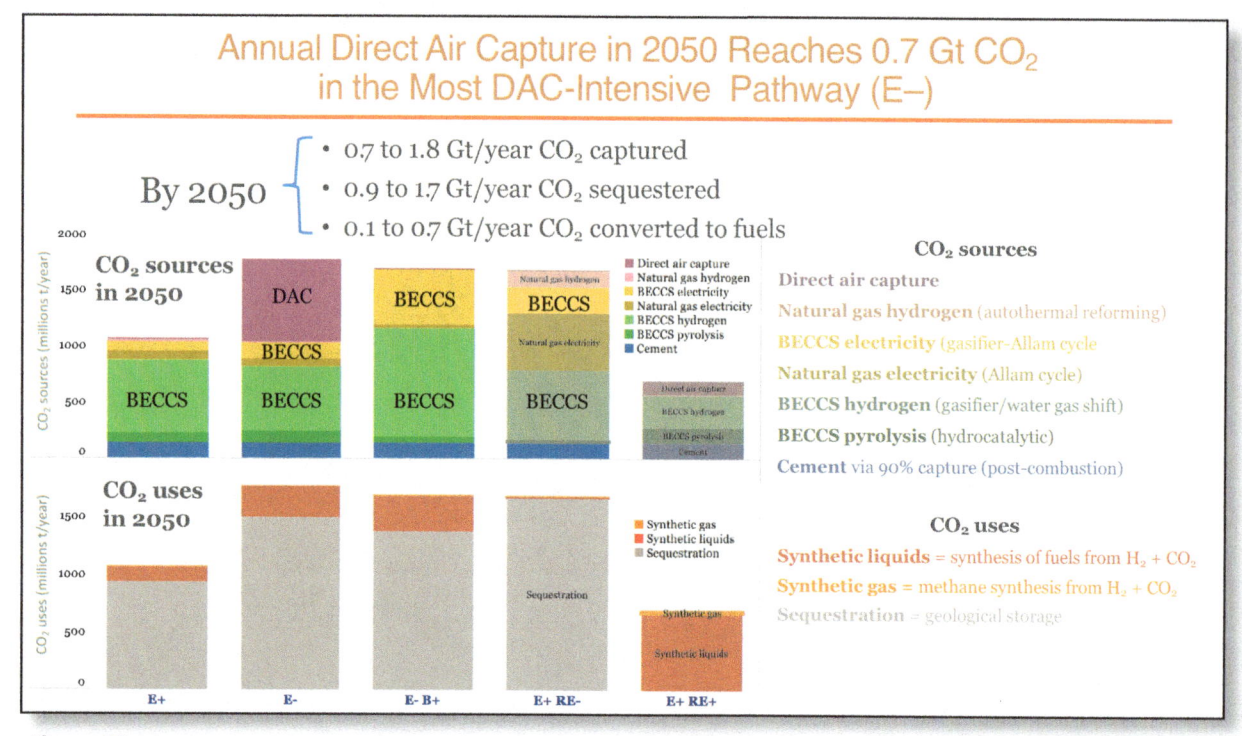

Figure 13

nologies play very important roles in the US in potentially getting to net-zero emissions—and potentially in other countries as well.

Just to recap what I have been talking about, why are we interested in negative emissions? Cumulative emissions of CO_2 determine future global warming. In order to stay below 1.5–2 °C of warming, the carbon budget that we have left to spend is shrinking quite rapidly. Negative emissions can essentially help us stay within our budget and meet those emissions thresholds.

What are the various net-zero emissions or negative emissions technologies? I reviewed a number of them, including restoring and managing terrestrial and aquatic ecosystems, mineralizing carbon, BECCS, and DAC with CO_2 storage. All of these have a role to play in meeting the carbon challenge. But BECCS and DAC have prospectively the largest roles. I showed you some results from our US study, including quite detailed modeling results, that really highlight the critical roles for those two particular future industries in reaching net-zero targets. Thank you very much for your attention. ◊

References

IPCC (Intergovernmental Panel on Climate Change). 2018. "Summary for Policymakers." In *Global Warming of 1.5°C: An IPCC Special Report on the Impacts of Global Warming of 1.5°C above Pre-industrial Levels and Related Global Greenhouse Gas Emission Pathways, in the Context of Strengthening the Global Response to the Threat of Climate Change, Sustainable Development, and Efforts to Eradicate Poverty* [Masson-Delmotte, V.P. Zhai, H.-O. Pörtner, D. Roberts, J. Skea, P.R. Shukla, A. Pirani, W. Moufouma-Okia, C. Péan, R. Pidcock, S. Connors, J.B.R. Matthews, Y. Chen, X. Zhou, M.I. Gomis, E. Lonnoy, T. Maycock, M. Tignor, and T. Waterfield (eds.)], 3–24. Cambridge, UK: Cambridge University Press. https://doi.org/10.1017/9781009157940.001.

Haszeldine, S., S. Flude, G. Johnson, and V. Scott. 2018. "Negative Emissions Technologies and Carbon Capture and Storage to Achieve the Paris Agreement Commitments." *Philosophical Transactions of the Royal Society A: Mathematical, Physical and Engineering Sciences* 376 (2119). https://doi.org/10.1098/rsta.2016.0447.

Larson, E., G. Fiorese, G. Liu, R.H. Williams, T.G. Kreutz, and S. Consonni. 2010. "Co-production of Synfuels and Electricity from Coal + Biomass with Zero Net Carbon Emissions: An Illinois Case Study." *Energy and Environmental Science* 3 (1): 28–42. https://doi.org/10.1039/B911529C.

McQueen, N., K. Vaz Gomes, C. McCormick, K. Blumanthal, M. Pisciotta, and J. Wilcox. 2021. "A Review of Direct Air Capture (DAC): Scaling Up Commercial Technologies and Innovating for the Future." *Progress in Energy* 3 (3): 032001. https://doi.org/10.1088/2516-1083/abf1ce.

Sanchez, D.L., G. Amador, J. Funk, and K.J. Mach. 2018. "Federal Research, Development, and Demonstration Priorities for Carbon Dioxide Removal in the United States." *Environmental Research Letters* 13 (1): 015005. https://doi.org/10.1088/1748-9326/aaa08f.

Session 1 | First Commentary
on the Presentation by Dr. Eric Larson

Cryogenic Carbon Capture as a Direct Air Capture Technology

Professor Larry L. Baxter
Chemical Engineering, Ira A. Fulton College of Engineering
Brigham Young University, Provo, Utah, USA

It is a pleasure and an honor to be part of the Twenty-Eighth International Conference on the Unity of the Sciences and to follow Dr. Larson in his discussion of net-negative carbon dioxide emissions. My name is Larry Baxter, professor of Chemical Engineering at Brigham Young University and cofounder of Sustainable Energy Solutions, which was recently acquired as part of Chart Industries to develop a carbon capture technology that I will speak briefly about in just a moment.

Sustainable Energy Solutions developed the cryogenic carbon capture™ (CCC) CO_2 mitigation technology to decrease costs and reduce the energy demand associated with CO_2 removal from point sources. How the cryogenic carbon capture process works is illustrated briefly in **Figure 1**. Flue gases from basically any continuous source enter the process and go through a heat exchanger to cool the gases to the point that CO_2 condenses from them. The CO_2 will typically desublimate to form a solid, and the solids and other gases are then separated.

The solids then warm by cooling the incoming gases, as do the gases, but they are now in separate streams. The solid

About the Speaker

- Distinguished Professor, Aalborg University, Denmark.
- Professor of Chemical Engineering, Brigham Young University, USA.
- Cofounder, Consultant, and Technical Director, Sustainable Energy Solutions (a Chart Company), USA.
- Researches in and develops carbon capture and sequestration and related climate change mitigation technologies.
- PhD in Chemical Engineering, Brigham Young University, USA.

Conceptually Simple Process

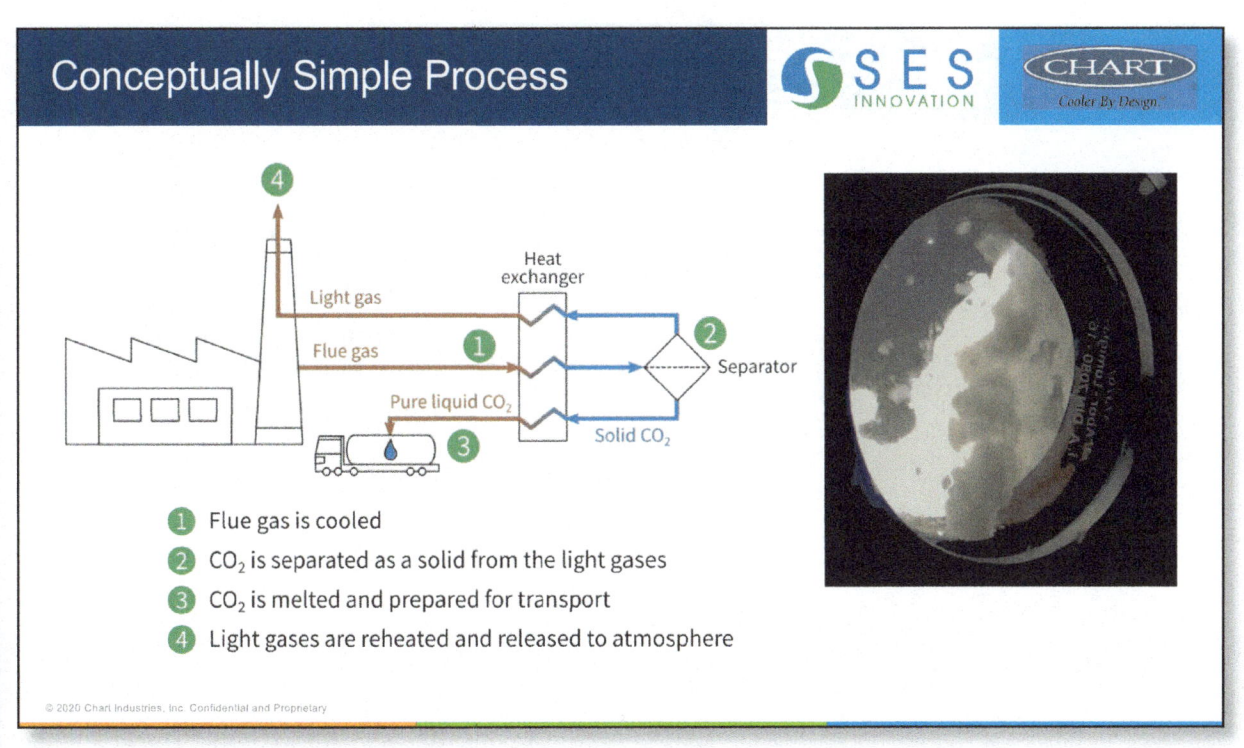

Figure 1

1. Flue gas is cooled
2. CO_2 is separated as a solid from the light gases
3. CO_2 is melted and prepared for transport
4. Light gases are reheated and released to atmosphere

stream is under pressure as it warms up and CO_2 melts. The image on the right in **Figure 1** depicts the solid CO_2 melting to form liquid CO_2 at about 10 bar pressure. The liquid is now further pressurized to form a pure liquid at 125 to 150 bar. This process significantly reduces the energy demand and the cost of carbon separation from point sources. However, the rest of my talk is going to be generally applicable to any carbon capture system and outlines what I believe to be a practical road map for net-negative emissions that is near term and reasonably affordable.

Carbon capture depends basically on three parameters: the purity of the CO_2 you produce, the fraction of the CO_2 to be captured, and the initial CO_2 content in the flue gas.

Figure 2 illustrates the theoretical minimum separation energy as a function of CO_2 product purity, ranging in this block from 95% to 100%. You can see that, as the purity increases, the theoretical minimum energy cost increases, but the energy cost increase is very small.

As you increase the CO_2 capture fraction, the minimum separation energy also increases. The increase is significantly larger than it was with purity. Over this range, you see it is still a fairly modest increase.

By far the largest sensitivity of the minimum separation energy for CO_2 capture is on the initial CO_2 mole fraction. **Figure 3** shows that dependence, and it swamps the other dependencies by orders of magnitude.

Direct air capture (DAC) is indicated

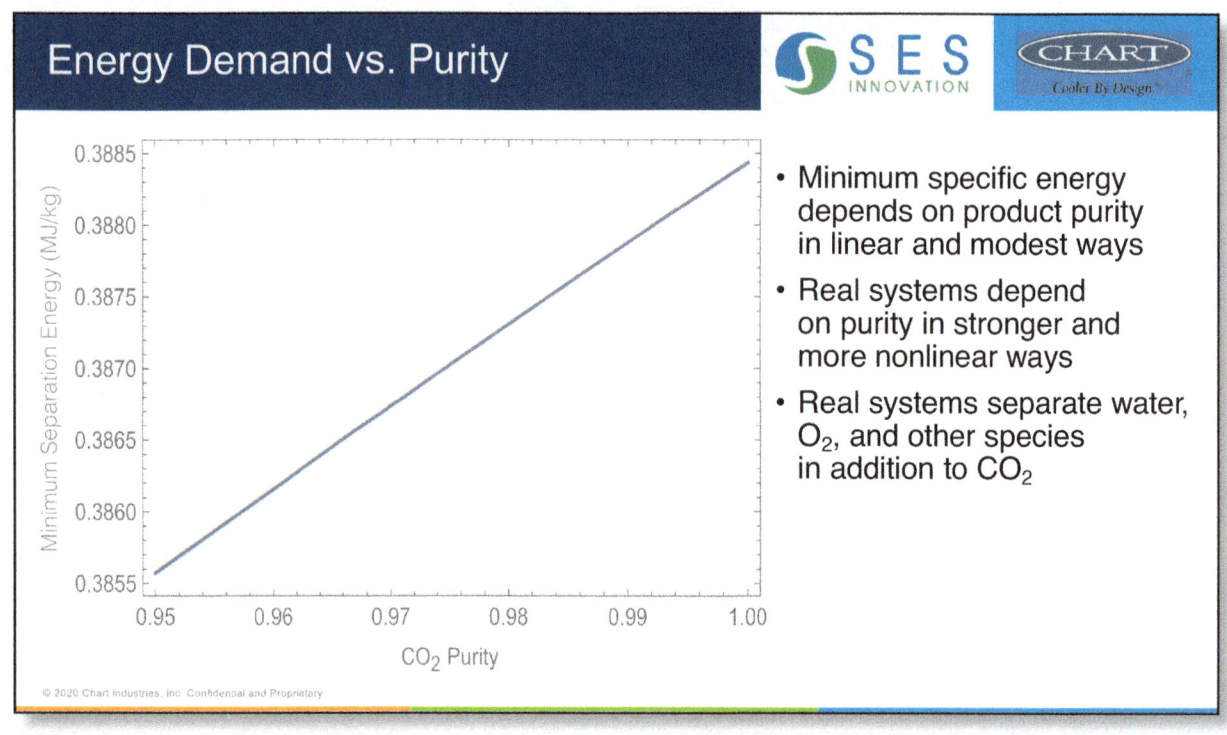

Figure 2

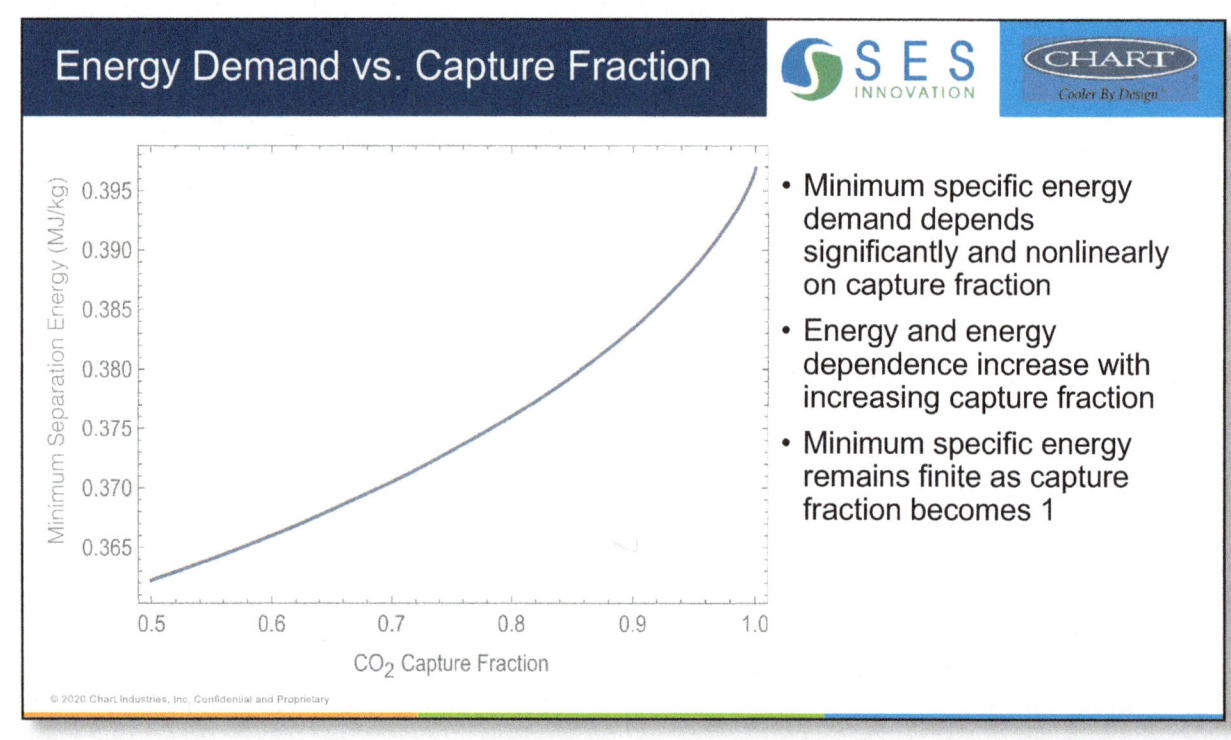

Figure 3

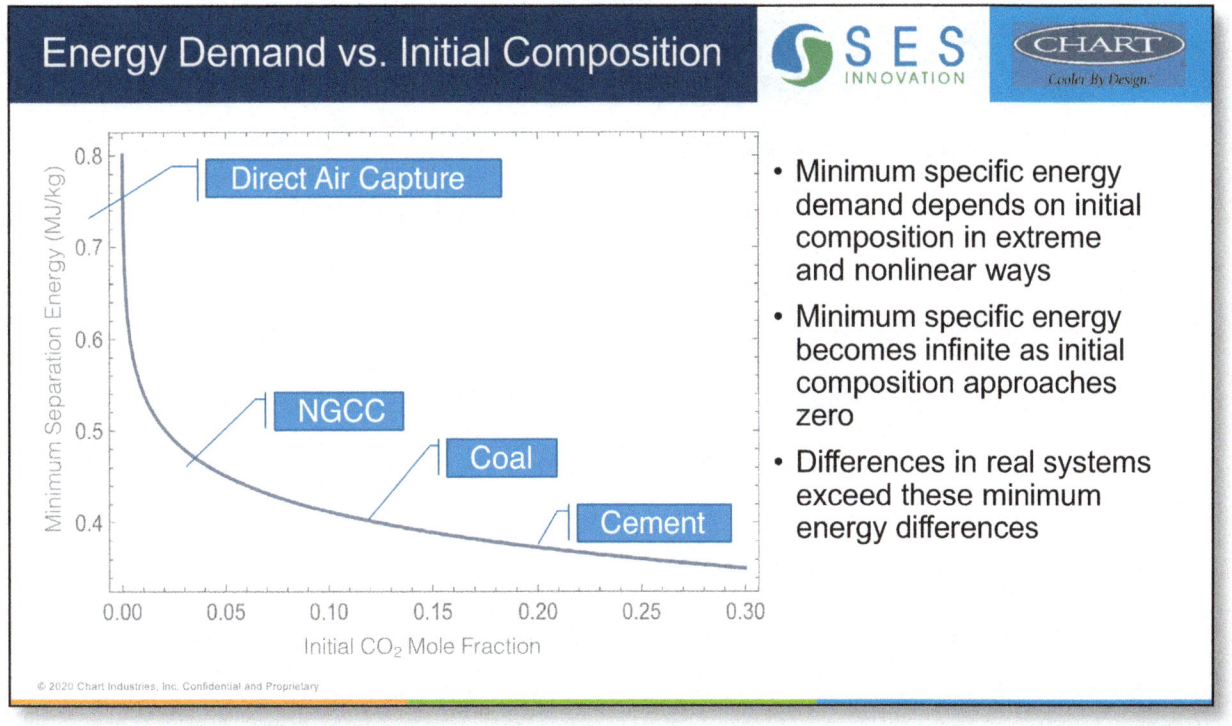

Figure 4

in **Figure 4** on the left, as it starts off with a little over 400 ppm CO_2. Typical natural gas combined cycle (NGCC) systems have 4–5% CO_2 in the gas, while coal-fired power plants have anywhere from 12–16% CO_2. And some industries, such as cement production, can have 23% or even up to 28% CO_2 in the flue gas. You can see that this initial CO_2 content has a major impact on the theoretical minimum energy that is required to remove a given fraction of CO_2.

While **Figure 4** shows the amount of energy required for both the separation of CO_2 from the other gases and the compression of CO_2 to 150 bar pressure, the plot in **Figure 5** looks at just the separation energy. Once the CO_2 is separated, the compression energy is similar for most carbon capture technologies (although the CCC separation process described above produces a liquid CO_2 product that has minimal required pressurization energy).

Again in **Figure 5**, you can see where DAC, natural gas combined cycle, coal, and cement fall out. For the separation energy, you can see that DAC, if you take the CO_2 directly from the air, requires a little over five times as much minimum theoretical energy as cement and coal and about four times as much energy as NGCC to capture the same fraction of the CO_2 source. What we are proposing here is that, rather than start with air as a feedstock, we implement point-source DAC systems by capturing all of the CO_2 that is produced from, for example, an NGCC

Point-Source DAC Advantages

- Uses existing gas-handling equipment
- Leverages point-source carbon capture equipment
- Operation and maintenance become incremental costs
- Transportation and storage infrastructure in place

Graph includes separation energy only, not including pressurization, etc.

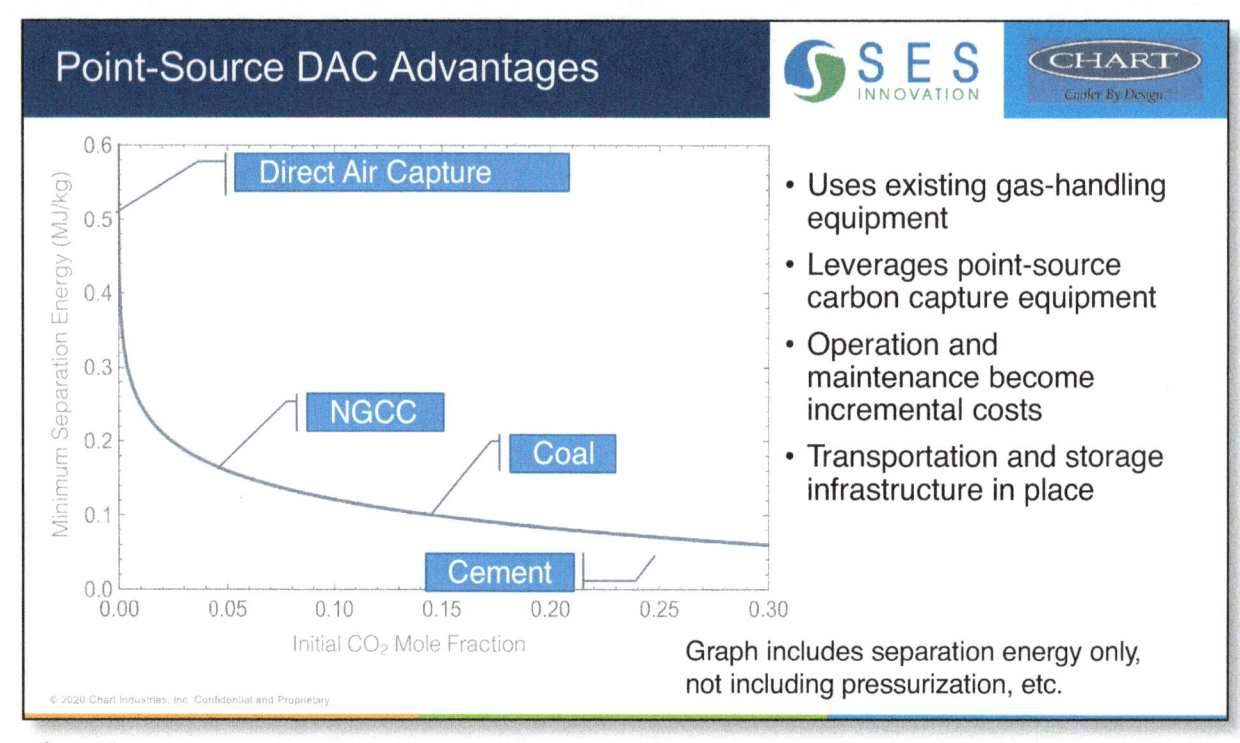

Figure 5

plant, a coal plant, or a cement plant, together with some of the CO_2 that enters those plants with the air that is used for combustion in those respective processes.

The quite large advantages of doing this are that, the gas-handling equipment from the existing process provides a reliable and high-volume gas stream and gas handling, which is the major capital expense and one of the major operating expenses for any carbon capture system. This approach also leverages the point-source carbon capture equipment. The capture fraction must be increased to greater than 99.7% capture in some cases, but the basic equipment only modestly changes to do this. Additionally, the operational and maintenance costs in this approach are only incrementally added to the already existing people and maintenance procedures for the original process whereas a dedicated direct air capture process must have dedicated resources of this type. Finally, the plant transportation and storage infrastructure that is in place for the first 99+% of the capture can easily manage the portion of the stream that constitutes direct air capture. Let us look at some quantitative results.

Figure 6 illustrates results from the CCC technology mentioned earlier, where the red horizontal line represents the threshold at which you cross into direct air capture. Operations below the red line have captured 100% of the CO_2 from the process and some fraction of the ambient air CO_2 that entered the process with the combustion air. As indicated, the process

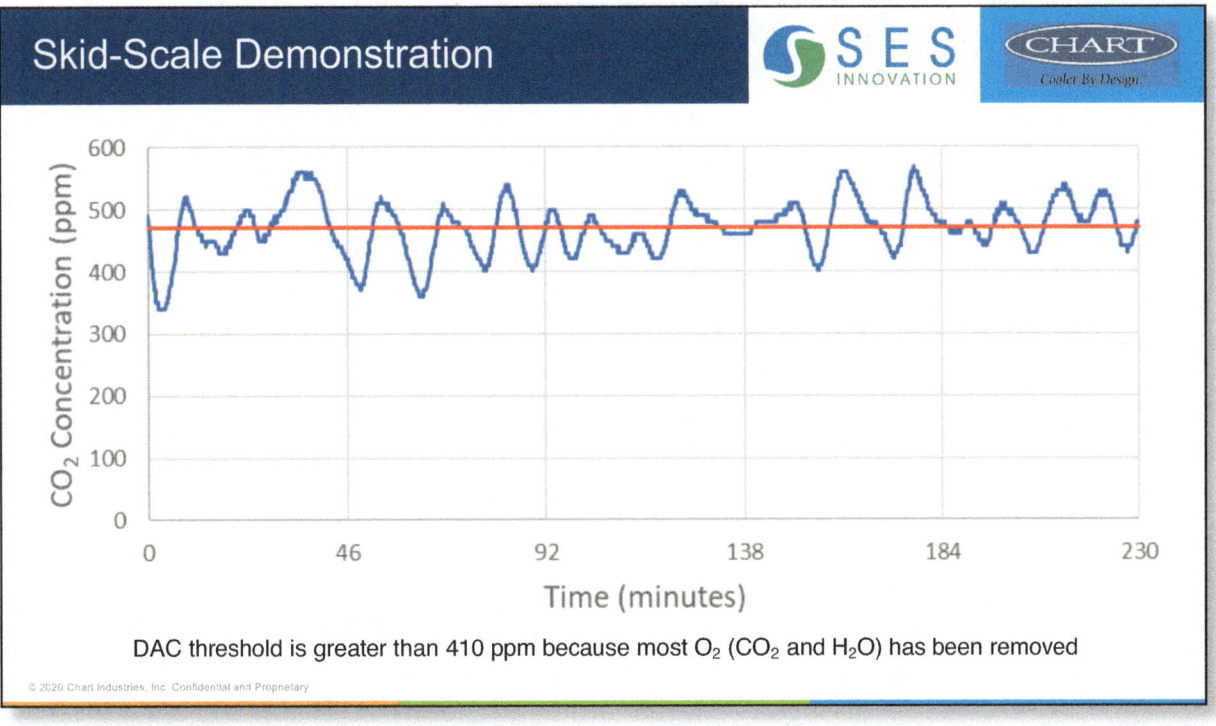

Figure 6

quite easily reaches the direct air capture point. (The CO_2 concentration at which you enter the direct air capture point is higher than ambient CO_2 concentrations because the process converts much of the oxygen in the air to CO_2 and water, both of which this process removes, which increases the effective concentration of the CO_2 that enters with air).

Figure 7 illustrates the effects of this type of direct air capture on cost and efficiency in terms of the capture cost in dollars per ton (left axis) and energy per ton of CO_2 captured (right axis). In this case, 100% capture represents capturing all of the CO_2 that the process produced, and the dots that are over 100% represent the inclusion of CO_2 that entered the process with the combustion air.

While the costs increase with increasing capture fraction, the increase is modest in terms of average costs and dollars per ton for CO_2 capture, even when you are capturing all of the CO_2 from the process and some of the CO_2 from the air. There is also a pretty incremental increase in the energy penalty associated with capturing atmospheric CO_2 in this way.

Some existing point-source systems (including CCC discussed above) can scale up capture percentages in this manner quite affordably while others struggle to do so. The CCC technology as currently developed and as being improved and commercialized can achieve such high capture percentages and provides a realistic, pragmatic way to actually get to net-negative

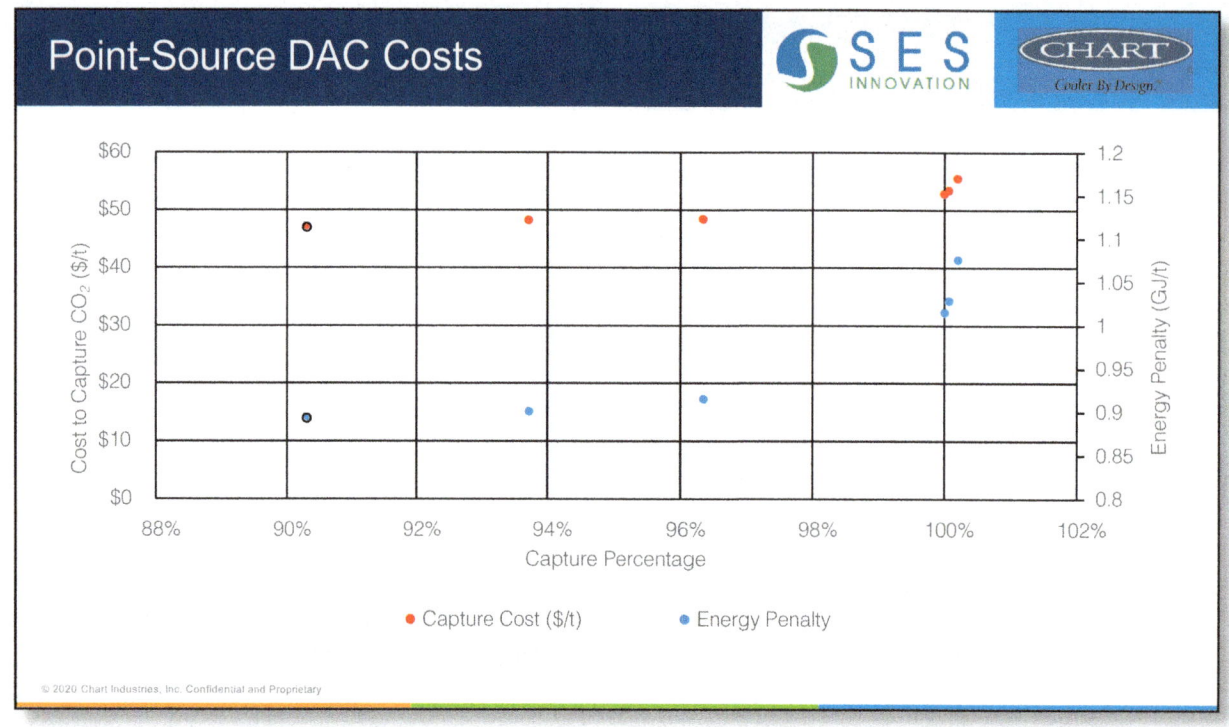

Figure 7

emissions. In addition, many of these sources can be fired with biomass to create processes such as Dr. Larson discussed, which will have even better energy efficiencies and cost numbers.

There are quite a number of sponsors, employees, and institutions that contributed to this work. I would like to acknowledge them at this point, and I look forward to the discussion. ◊

Session 1 | Second Commentary
on the Presentation by Dr. Eric Larson

Negative Emissions Technologies in U.S. Decarbonization Pathways

Professor Steven S.C. Chuang
Polymer Science, School of Polymer Science and Polymer Engineering,
The University of Akron, Ohio, USA

Dr. Larson's presentation outlined the six essential pillars of decarbonization: 1) end-use energy efficiency and electrification, 2) clean electricity, 3) clean fuels, 4) CO_2 capture and utilization or storage (CCUS), 5) reduced non-CO_2 emissions, and 6) enhanced land sinks. These six serve as excellent guideposts for policy makers and researchers toward further development of cost-effective and energy-efficient technologies.

Bioenergy technology with CO_2 capture and underground storage (BECCS), discussed in **Figures 5–7** of his paper, is highly promising. However, there is a need to point out its limitations due to the availability of biomass (its growth, harvest, collection, transportation, and purification) and the related CO_2 emissions from its use. The use of bioenergy, such as biodiesel or biomass, strongly depends on the geological location. Other forms of renewable energy—wind, solar, and hydroelectric power—could also be effective in achieving a negative carbon footprint as well as in minimizing the overall environmental impact when integrating with CCUS.

The global and US trend in CO_2 emissions is clearly not on the right track despite the many discussions, global cli-

About the Speaker

- Professor of Polymer Science, The University of Akron, USA.
- Director, FirstEnergy Advanced Energy Research Center.
- Researcher in clean energy generation and storage, including the fabrication of polymeric and hybrid composite materials for carbon dioxide capture, proton-exchange membrane/solid oxide fuel cells, and photo/electrochemical processes for the utilization of solar energy.
- PhD in Chemical Engineering, University of Pittsburgh, USA.

mate agreements, and intensive research and development efforts. The challenge in transitioning from a fossil fuel–driven society to a carbonless and carbon-neutral one is daunting. The most formidable barrier in the transition to a carbon-neutral society can be understood to be the significant difference in the energy/power density, cost, and availability of fossil fuels compared to those of renewable energy. The order-of-magnitude difference in power density between fossil and renewable energies can be traced to basic chemical kinetics: the Arrhenius equation, $k = e^{-E/RT}$. It shows that the rate of energy generation of fossil fuels (that is, the power density of their combustion at high temperature) is often at least one or often several orders of magnitude higher than that of renewable energy sources at ambient temperature. It is important to note that Savante Arrhenius, who proposed the Arrhenius equation in 1889, was the first to determine the contribution of CO_2 to the greenhouse gas effect.

Many of the current renewable energy technologies, including bioenergy, have a low efficiency and are not cost effective. Therefore, the enhancement of end-use energy efficiency and the other pointers in Dr. Larson's six pillars, as well as the efficiency of those steps involved in the overall process of energy generation, conversion, and usage, are essential for moving toward a carbon-neutral and carbon-negative society. The latter requires successful deployment of cost-effective CCUS technologies.

As pointed out in **Figure 7** of Dr. Larson's talk, CO_2 capture technologies are well understood but are not widely deployed. I would like to elaborate on the issues regarding the efficiency of the current CO_2 capture technology and briefly discuss the barriers for its further development and deployment. The current large-scale Climeworks CO_2 capture process and the planned 1PointFive process employ a thermal swing approach, with an amine-based sorbent and a K_2CO_3 sorbent, respectively. The temperature swing adsorption (TSA) with amine-based sorbents is operated by the adsorption of CO_2 at low temperature, typically below 60 °C and regeneration of the sorbent or solvent at temperatures above 100 °C with a dual adsorber—one for adsorption and the other for sorbent regeneration. The barrier for the wide adoption of TSA, including many proposed CO_2-negative emissions technologies, is their low energy efficiency, which translates to high operational cost.

Further improvement in the efficiency of the TSA process requires detailed analysis and integration at each level: 1) the structure of amine sites and their neighboring functional groups at a nanometer (that is, molecular-level) scale in the sorbent particle, 2) incorporation of the amine sorbent particles into the internal structure of the adsorber at a centimeter and meter scale, 3) the mode of the adsorber operation—whether TSA, pressure swing adsorption, vacuum swing adsorption, or vacuum-assisted thermal swing adsorption (VTSA), and 4) the composition of the CO_2-containing feed stream.

At the molecular level, a fundamental understanding of amine–CO_2 chemistry is needed to advance to a level for manipulating the chemical environment of amine sites to reduce their binding energy with CO_2 and to enhance the rate of both adsorption and desorption. Reduction in binding energy and enhancement in adsorption/desorption without affecting CO_2 capture capacity will substantially decrease the energy penalty, allowing the TSA process to be operated at low sorbent regeneration temperatures and prolonging the operational life of the amine-based sorbents. At the centimeter/meter scale, improving the heat transfer of the sorbent structure will decrease the cycle time of the TSA process, thus enhancing the efficiency of the overall process.

The mode of operation, which is governed by the structure of the adsorbers and the sorbents as well as the packing of the sorbents, has a significant impact on the purity of the captured CO_2. A cost-effective, energy-efficient CO_2 capture process must be capable of generating a high purity of captured CO_2 to meet the purity requirement for CO_2 sequestration and utilization without additional separation processes. We recently found that VTSA is effective in producing a high purity of captured CO_2 ("Gradient Amine Sorbents for Low Vacuum Swing CO_2 Capture at Ambient Temperature," DE-FE0031958, 1/1/2020–6/30/2022). VTSA also shortened the cycle time (that is, the time needed for a VTSA cycle including four steps: CO_2 adsorption, heating, blowdown, and evacuation).

I concur with Dr. Larson that both BECCS and direct air capture with carbon storage are technically feasible and highly promising. Dr. Larson's presentation highlighted the key features of negative emission technologies (NETs) and provided a clear picture on how NETs could pave the way for the US to achieve net-zero emissions by 2050. ◊

Session 1
Discussion

Cliff Davidson (Chair): As we enter the discussion period, for any participant who would like to ask a question, we ask that you click the button at the bottom of the Zoom window where it says *Raise hand*. Then I will acknowledge you. Also, when you begin your question, it would be good to say which presenter or commentator your question is addressed to, if there is a specific individual. With that, I will ask if there are any questions for our speakers. Okay, I see Dr. Thomas Valone. You may go first. Thank you.

Thomas Valone: This question is for Dr. Larson. We are all talking about direct air capture, which is very laudable. We are also praising the negative emissions goal by 2050, which is also worthwhile. However, in my papers in half a dozen journal articles that I have published, including one in the chat, you can take a look at the issue that we are addressing in our institute, that being the excess 830 Gt of CO_2 in the atmosphere now, with basically the argument that it is contributing the most to global warming. When we consider preindustrial levels, up to 290 ppm of carbon dioxide was really the maximum we had seen for 400,000 years, thanks to Dr. Jim Hansen (NASA climatologist), who published a climate graph in *Technology Review* way back in 2006. We are now up to 410 ppm worldwide. When we take the difference between those two numbers, that's how you get the 830 Gt of excess carbon dioxide. Therefore, my question is, will the Larson team address direct air capture on the gigaton level? There are a few companies that are actually looking at that and considering the reality of removing the 830 Gt over the next few decades. I was wondering if he has looked at that himself.

Eric Larson: Thank you for the question. We have not explicitly looked at that issue. I think you make a really good point that the amount of CO_2 that we have put into the atmosphere since preindustrial times is very significant. That said, the modeling work that we did does reach on the order of 1–2 Gt of CO_2 removal each year, just for the US, by 2050, and that would need to essentially continue for quite some time. Imagining that kind of a system to actually be developed in the US is difficult for me because that is such a new direction from where the country has been going. On top of that, looking at removing the cumulative 830 Gt worldwide that you mentioned is an even more daunting task. Thank you.

Davidson: Thanks for your question, Dr. Valone. Next, we will have a question from Staffan Berg.

Staffan Berg: Yes, thank you very much for allowing me to comment and ask Dr. Blekhman a question. I am so excited to be in the middle of this because, in my youth, I went to Chalmers University of Technology, which Dr. Blekhman also visited, and I studied chemical engineering, which is what this seems to be about. But my question is about electric vehicles and why they get so much press coverage as compared to fuel cell vehicles or hydrogen vehicles. The latter technologies do not seem to get much media attention at all. What is it that is holding back this hydrogen or methanol technology?

David Blekhman: Thank you very much for your wonderful question and flashback to Chalmers. As a matter of fact, Chalmers University of Technology is very actively pursuing electrification through their collaborations with Volvo and Scania. There is a wonderful electrical vehicle center where Chalmers is a key player and the host on campus. There is thus a very strong effort.

First of all, both electric vehicles or battery electric vehicles and fuel cell electric vehicles are clean transportation, and therefore they have a valid place in the market by providing us with zero-emission technology. In addition, batteries are a relatively mature technology in many ways and provide sufficient transportation for our commuting needs. If you look at a typical American commute, you can do it in under 40 miles round trip.

Thus, a battery-powered vehicle can suffice for everyday transportation needs and is thus popular. For relatively short distances, you can even charge such vehicles overnight at home or for a few hours at work, and you can get 100 miles. However, if you go to high-power battery vehicles, you get a much longer range but at the cost of very demanding and difficult charging. Long distances present a challenge. The charging time is also very significant, or the charging infrastructure is very expensive per single vehicle if you want to reach charging rates of 50 kWh/h or 100 kWh/h. A typical Tesla will be around a 75–100-kWh capacity that you would need to put into this vehicle. Thus, for the majority of our needs, we can actually handle transportation.

This is a strong alternative, but I have a general problem with lithium versus hydrogen technologies in terms of the limited access to batteries. Only a few certain countries mine lithium, and few certain industries process lithium, so access to wealth generation through batteries is limited for electric vehicles. Otherwise, batteries are a wonderful technology. Right now, we are looking at using hydrogen not for passenger vehicles but for heavy transportation that will be traveling 500 miles or over 500 km or perhaps 800 km per day. That is where hydrogen starts to clearly win the game over electric vehicles. After introducing hydrogen for heavy-duty transportation, we are hoping for light-duty long-distance vehicles. Fuel cell passenger vehicles will play a more significant role. But overall, just on the passenger side, fuel cell vehicles are very wonderful, able to travel very long distances cleanly with no emissions—just water vapor.

In addition, with these vehicles you can refuel in under 10 minutes, even 5 minutes, and then you can go 400 miles. There are clear advantages. However, not every consumer needs a fuel cell vehicle with that capability and can get away with an electric vehicle. This is still playing out in the marketplace. Right now, electric vehicles are more popular because they are easier to handle and the infrastructure is developing rapidly. But with heavy-duty transportation advancing, we are hoping that the passenger market will catch up. Also, there is another topic, that being forklifts, and I would encourage you to explore hydrogen forklifts and how they are starting to play a significant role in the logistics industry. That is an interesting story in itself. Thank you.

Davidson: Excellent, thank you very much. And thank you, Staffan Berg, for your question. We have another question now from Hiroko Oizumi.

Hiroko Oizumi: Thank you very much. My question is for Prof. Blekhman and is very much related to the previous one. It requires just a simple answer. Toyota, Japan's leading car company, started producing hydrogen-powered cars quite a long time ago. However, EVs—without hydrogen, of course—have been prevailing throughout the world. Thus, in my observation, hydrogen-powered cars have been making slow progress. But perhaps due to the Russian invasion of Ukraine, things might have changed. Hydrogen and nuclear power have become more important. Hydrogen cars and hydrogen power have a chance to be developed now. My question is very simple: Which will prevail in the market, hydrogen-powered cars or EVs? I know that Prof. Domen said that hydrogen is very expensive, and Dr. Edlund said that marketing is very slow. Therefore, perhaps it is up to the government to choose hydrogen rather than anything else in order to promote the technology—whether to promote this hydrogen technology or not. This of course falls in the realm of politics. I am not a scientist but a politician. Thank you.

Blekhman: It is indeed a simple question. But the answer is actually quite complex. Thank you for thinking so broadly about hydrogen technologies.

First of all, I do not think the Russia–Ukraine war is a cause here. We have been looking at clean technologies for a long time and trying to refine our approaches to climate change and clean air. The European Union has been doing really good work in introducing such approaches, and I really would like us to stay away from just directly focusing on hydrogen vehicles or electric vehicles. What we are observing now, and what we have come to realize, is that with the introduction of renewable energy like solar and wind, we need intermediate energy storage in transitioning from fossil fuels. There is another driver for renewable energy: the use of fossil fuels would not last forever.

Even though we are still finding more and more fossil fuel deposits, the reality is that they are a finite resource and they result in CO_2 emissions, making them a double-edged sword. In reality, we are addressing a global problem with more intricacies and more system elements than just vehicular emissions. The energy consumption in any country is four-pronged: commercial, residential, industrial, and transportation. We are trying to address all of those, globally.

In the United States, these prongs are equally divided, more or less. Thus, transportation accounts for only one-quarter or so. If you look at heavy-duty transportation, such as trucks, shipping, and aviation, they are all contributing—not just cars—so we are solving global problems and going back to renewable resources. Now we are going to store that energy, and there are not enough batteries in the world to store that much energy, whereas storing it in hydrogen is a much more efficient process. This is because we are storing only 1 kg of hydrogen and releasing 8 kg of oxygen, so we do not have to deal with the weight of oxygen. If you are storing it in the battery, there is so much chemistry and weight in things involved in these processes. What I am saying, then, is that we are solving a lot of global problems. If you look at my presentation, I am really looking at this global picture of creating hydrogen hubs in large systems that are more synergistic in resolving energy issues. Thank you.

Davidson: Great, thank you very much for that question. And thanks for your response. Okay, we have a question from Takahiro Hiroi.

Takahiro Hiroi: Thank you. My question—perhaps half question and half comment—is regarding the very fundamental topic of this session. I am a planetary scientist. Over the last 4.55 billion years of history, the Earth has reduced the amount of CO_2 to the minimum level at this moment. Also, temperature-wise, it is really kind of a small ice age right now; it is not so warm. The Earth has its own natural system to adjust its environment. I wonder whether undertaking these artificial ways of achieving carbon neutrality, such as by weathering on the land or whatever, which are actually done very slowly in Nature, with humankind's very short-term knowledge and experience could be dangerous and could disturb Earth's balance.

For example, we used to believe that not cutting trees was good for the Earth. But actually, we have to cut trees to create a healthy forest cycle and use that accumulated carbon for other uses instead of erecting steel buildings. We learned a lesson from our mistakes the hard way. Therefore, I would be much more cautious about trying to control CO_2 levels artificially, because doing that still has a carbon footprint, such as from constructing facilities and operating things. I am more inclined to use natural methods, like cutting trees or planting more bamboo forests. I may be a little biased, but it is my opinion.

Davidson: Who is your comment directed toward?

Hiroi: It is not directed toward anyone in particular. It is just regarding the fundamental assumption of this session to reduce carbon dioxide.

Davidson: Okay. Well, let me ask if any of our speakers would like to give a response to this point.

Larry Baxter: I will respond. While the Earth has seen a lot of variation in climate and has adjusted to those variations, it is not the case that the Earth has been warmer during human civilization. Another major issue in addition to the level is the pace at which the CO_2 is changing. That is what is going on. In the past, through most natural changes, soil types and other things have adjusted slowly as the Earth's temperature varied. We are now moving things much more rapidly than they moved during these natural fluctuations. Thus, rather than a slow and measured approach, the science, including the report that came out just last week, is pointing to great urgency on this issue and that we really need to be very aggressive in trying to manage it.

Davidson: Great, thanks.

Hiroi: Thank you.

Davidson: Okay. I guess there will be time for further discussion. We do have another question from Dr. Valone.

Valone: I wanted to not only respond to Mr. Takahiro's comment, but also ask Dr. Larson about the carbon budget that he was referring to, which seems quite arbitrary and only allows for more fossil fuel usage. But before I do that, may I show my screen? It requires your approval and enabling instead of disabling my Share Screen button.

Davidson: I am sure you could share your screen if the people in control of the screen can do that.

Valone: The host has disabled it at this time. But yes, if you allow me to press this Share Screen button, I can show the past 400,000 years of carbon dioxide or temperature and sea level, thanks to perhaps the best climatologist in the world, Dr. Jim Hansen.

Davidson: Yes, I am just getting a message from the individuals who are controlling the technology here that it is not going to be possible to do Share Screen.

Valone: Then I will just go ahead with the question. If people want to see the graph I was about to show, they can do that by looking in the chat window and clicking on the link for the climate chart; it is a .pdf file. My question to Dr. Larson about the carbon budget is that we are supposedly allowed to somehow maintain 2 °C of increase? I am very curious and confused about that calculation.

Larson: Yes, that is straight out of the IPCC's reporting that looked at the historical data on CO_2 emissions since the mid-1800s. We know pretty well what the cumulative emissions have been, and we have measured the temperature rise that has gone along with that. That is basically the relationship. I am not a climate scientist, but I trust the analysis and the scientists who put that together, which shows quite clearly that the warming that we will experience—even if it has not risen all the way to the final level of warming—we can expect to be correlated with the cumulative emissions. Maybe if any of my cospeakers here want to add to that, feel free to do so.

Valone: To give me a list of emissions is probably key there. Maybe the IPCC was looking at something else. But sure, it would be appreciated if anyone else wants to answer the question. Anybody?

Baxter: I will make a comment that it seems that too often we get into an argument about what technology is the best or what our goal should be. The reality is, we have accomplished precious little thus far in terms of control, managing the climate, and any technology that can realistically have an effect. There is plenty of CO_2 for all technologies and approaches. Your argument might be that we should not think about the budget leftover. I think the more pertinent point is that we need to start doing more and putting more things on the ground that actually make a difference. We do need to think critically about whether those technologies are efficient or effective, but if they are effective and efficient, there is certainly room for lots of options in that they are relatively urgently needed.

Davidson: Right. Thank you very much, Dr. Baxter. And let us see, does anyone else want to address Dr. Valone's question? Or should we move on? Okay, I do not think there is anyone else. I have a question for Dr. Larson. It seems to me that this is really the idea behind the negative emissions, that if you have got just $1 to spend, you are going to be so much more efficient at removing CO_2 from, say, stack gases, where the concentrations are very high compared to trying to remove CO_2 from the atmosphere, where you have an incredibly high dilution factor. And I am just wondering if you could talk a little bit about comparing the cost and how that is related to the physics of those two different processes.

Larson: Yes, the cost, you can say, is coarsely proportional to the concentration of the CO_2 that is in the stream you are trying to remove it from. One of the commentators I think showed that—as did the energy-system cost resulting from our modeling, which was a least-cost optimization for the energy system. The reason that the direct air capture did not show up as significantly as other capture alternatives, in particular the biomass option, is because the cost is higher for the direct air capture. We have not done enough direct air capture yet to know really what the real costs are going to be at this point. But the best engineering estimates that have been made for what it would cost to build these various devices that have been created or imagined say that the direct air capture option is going to be more expensive, as you pointedly put it, than capturing from flue gases. Larry, do you want to add anything to that?

Baxter: Just to point out, the incremental cost of taking a point source emission such as from an exhaust stack extending it to direct air capture levels, is quite significantly lower than the cost of purpose-built direct air capture systems. Both the energy and the cost are quite a bit lower, and the marginal cost of capturing the last little bit of CO_2 is still even then on the order of a couple hundred dollars per ton. But that is probably a third or a fourth that of a purpose-built system.

Davidson: Thanks very much, Dr. Baxter and Dr. Larson. Okay. Are there any other questions or any other comments that people would like to raise before we close the session?

Well, I am hearing no more questions. I would like to thank all of our presenters and all of the commentators, as well as those who asked questions afterward. I think this has been a most interesting session, and certainly a lot of important points have been raised and debated. Ah, now, is there a question from Jessica Guichard?

Jessica Guichard: Ah, yes. I am very excited to participate in this conference. I just started working on a project where we are looking at producing hydrogen using offshore renewable energy. In light of this, I wanted to comment on the difference between electric vehicles and fuel cell–powered vehicles, which some people asked about. For example, I recently worked on a fuel cell–powered boat where it actually needed both batteries and fuel cells—you did not just have fuel cells in the boat, so it was a bit more complex than just electric batteries. But one advantage of fuel cell–powered vehicles—which are not developed as much—is that hydrogen can be stored for a long time, whereas with electricity, you have to use it up. For example, if you produce excess electricity over a certain period of time, it is a very good idea to use this electricity to produce hydrogen. Therefore, being able to have vehicles that use hydrogen will be good when there is more and more renewable energy, I think. Yes, so that is just my thought.

Also right now, actually, most of the hydrogen produced, as one of the commentators said, is still gray hydrogen. It is not being produced using electrolysis yet. Actually, unfortunately, people who already have fuel cell–powered vehicles will not be using clean energy for power most of the time, at least not for the source of the hydrogen. So, these are my comments.

Davidson: Right. Thank you very much, Dr. Guichard. Okay, I think we are now just about at the closing time. Once again, I want to thank all of the speakers and the questioners, and I look forward to further great discussions in this conference. Thank you all. ◊

ICUS XXVIII

SESSION 2

Manufacturing Materials for Eco-Friendly Products

About Session 2

Manufacturing Materials for Eco-Friendly Products

Advances in chemistry and chemical technologies have made it feasible to design and manufacture a vast array of materials and products that are useful for everyday life–including products we use for food and clothing, dwelling places and workplaces, travel and shipping, health and medicine, arts and entertainment, and more. As the human population continues to grow, the demand for goods and services also increases. At the same time, problems associated with the use of toxic materials and the disposal of wastes need to be resolved. This session outlines these challenges and offers ways to address them.

The first presentation discusses the methods, applications, and advantages of "green chemistry" for sustainable development. Green chemistry involves the use of eco-friendly (nontoxic) substances and processes, it is geared toward improving efficiency in the use of energy and material resources, and it is coupled with the appropriate degradation or recycling of wastes.

In the second presentation, the focus is on plastics. It considers the pros and cons of various types of plastics, and it discusses the use of renewable feedstocks and the production of biodegradable plastics. It places the challenges of plastic sustainability within the larger context of environmental sustainability. ◊

About the Chair, Professor Michael Stenstrom

- Distinguished Professor, Department of Civil and Environmental Engineering, University of California, Los Angeles, USA.
- Former Chair, Department of Civil and Environmental Engineering, University of California, Los Angeles, USA.
- Research in process development for stormwater management and wastewater treatment systems.
- PhD in Environmental Systems Engineering, Clemson University, USA.

Session 2 | Presentation One
The Promise of Green Chemistry

Hon. Mike Lancaster
Head of Innovation,
Chemical Industries Association, London, UK

This paper looks at some of the successes, drivers, challenges, and issues surrounding the application of *green chemistry*, particularly in the chemical industry, and at what must be achieved in the future to meet current global challenges. This is done with reference to three of the UN Sustainable Development Goals—namely, management of water supplies, sustainable production and growth, and climate action—which a growing number of companies in the chemical industry are signing up for. It is argued that the pace of development of more sustainable products and processes needs to accelerate, and that underpinning this is a requirement for even faster expansion in the teaching of green chemistry, such that all chemistry becomes green chemistry.

Introduction

The term *green chemistry* was first used in the early 1990s, and early efforts are widely viewed as a direct response to the US Pollution Prevention Act of 1990. The concept that developed during the 1990s culminated in the formulation of the 12 Principles of Green Chemistry (**Figure 1**) by Paul Anastas and John Warner in 1998. This sparked interest across the globe, particularly in academia, with new master's and undergraduate programs

About the Speaker

- Head of Innovation and Events, Chemical Industries Association, UK.
- Former Green Chemistry Network Manager, University of York, UK.
- Former Business General Manager, Electroplating Chemicals, Yorkshire Chemicals, UK.
- Former Senior Project Leader, Specialties and R&D, BP.
- MPhil, Chemistry, University of Leicester, UK.

The 12 Principles of Green Chemistry

1. Prevention
2. Atom Economy
3. Less Hazardous Chemical Synthesis
4. Designing Safer Chemicals
5. Safer Solvents and Auxiliaries
6. Design for Energy Efficiency
7. Use of Renewable Feedstocks
8. Reduce Derivatives
9. Catalysis
10. Design for Degradation
11. Real-time Analysis for Pollution Prevention
12. Inherently Safer Chemistry for Accident Prevention

Figure 1

Some Barriers and Drivers for the Adoption of Green Chemistry

Barriers	Drivers
Lack of global harmonisation in environmental regulation	Legislation driving innovation
Green Chemistry teaching and research still somewhat specialist rather than the mainstream.	Growing public concern over environmental issues
Short term drive for profits	Better relationships / reputation with stakeholders, regulators, and the public
Insufficient funding and short-term paybacks required by industry	Customer / consumer demand
Relatively inexpensive well developed fossil fuel / raw material supply chain	Sustainability metrics becoming well developed
Lack of universally accepted metrics	Potential to help solve global problems e.g. climate change
Focus on exposure prevention rather than risk reduction	Commitment to sustainability goals

Figure 2

being developed. Awards for green chemistry research and for the commercialization of greener products and processes also began to be issued.

In the 30 years since green chemistry's inception, much progress has been made in the "invention, design, development, and application of chemical products and processes to reduce or to eliminate the use and generation of substances hazardous to human health and environment" (Anastas and Warner 1998). However, success has arguably proven to be more evolutionary than revolutionary. If we are to solve some of the major problems facing society today, the pace of progress needs to accelerate over the next 30 years.

Globally, the knowledge, adoption, and teaching of green chemistry are now quite well established, and there are dedicated green chemistry courses at the graduate and postgraduate levels in many countries. Also, several large chemical and pharmaceutical companies have adopted the green chemistry philosophy and developed metrics to measure progress. However, the green chemistry system is still far from being *the* way in which chemistry is taught and carried out in laboratories and production facilities worldwide. Fairly new developments such as the Green Chemistry Commitment from Beyond Benign (https://www.beyondbenign.org/) are a great idea to improve the situation.

Barriers and drivers

Before looking at the promise of what can be achieved, it is worth studying some of the barriers and drivers that, respectively, have hindered and to some degree propelled the adoption of green chemistry (**Figure 2**). We need to remove these barriers and embrace the opportunities offered.

Legislation has proven to be both a barrier and a driver. There has been exponential growth in environmental and related regulations aimed at ensuring that the products of and waste from the chemical and pharmaceutical industries minimize harm to both the environment and people. In Europe, far-reaching examples include the REACH regulations (Registration, Evaluation, Authorization, and Restriction of Chemicals) and the Industrial Emissions Directive.

Most countries now have or are developing similar regulations, but we are far from global harmonization on environmental regulation of this kind. These regulations have come at a huge cost for the sector, and some less scrupulous companies, with a focus on short-term profit, have chosen to manufacture in locations with limited environmental controls and to continue with chemistries that fail to rein in pollution and waste. However, for more responsible companies, these regulations have stimulated innovation and more sustainable product and process development. In a global economy, we need an even regulatory playing field.

As just noted, there has been a proliferation of green chemistry courses and modules, but the green approach has yet to become the default way of working, even for recently graduated chemists. Indeed, standard practice in many industrial labs mitigates against green chemistry. Just look at the way we input energy

Figure 3   **Source:** sdgs.un.org/goals

into reactions. Green chemistry teaches that heating a whole reactor or flask of organic solvent with some highly diluted reactants therein is not always the best procedure. Microwaves, ultrasound, photochemistry, mechanical grinding, and so on are sometimes more efficient ways to impart energy where it is required—but how many laboratory benches have such equipment readily available? Very few. We still rely predominantly on electrical heating mantles and oil or water baths. We cannot hope to develop green commercial processes if we do not equip the next generation of chemists with the appropriate knowledge, research tools, and techniques.

In developed economies, certain characteristics of the chemical industry also make commercial implementation of green chemistry more problematic. Many of the processes operated and equipment used are well over 30 years old and fully discounted. New, innovative procedures are being developed, along with more efficient equipment, but the cost of replacing old plants or processes with these new ones may be too much for a business to handle (unless the company was going to build a new plant anyway). Many major companies demand a payback period of two years or less in order to approve new capital equipment. This undoubtedly delays the introduction of newer, sustainable technologies. To overcome this hurdle, companies need a longer-term vision and strong leadership. Appropriate policy and regulatory drivers will also help, as will access to financing and government grants.

Despite the barriers, progress continues to be made as climate change, protection of the environment, and the wider sustainability agenda become of ever-increasing importance to the public, governments, and global organizations. Many companies are now signed up with the UN Sustainable Development Goals (SDGs) (**Figure 3**) and are using the Science Based Targets initiative. These commitments, often demanded by stakeholders and the market, along with the associated auditing and public disclosure, will be key drivers for the future.

To look in more depth at the specific role green chemistry must play in helping to solve the major problems facing the world today, I have chosen to consider its potential within the context of three of the UN's SDGs. These goals have become one of the key means for driving and measuring progress across the world.

Goal 6: Ensure availability and sustainable management of water and sanitation for all

It is widely acknowledged that chlorination of drinking water has saved many millions of lives. However, according to the UN, 2.2 billion people still lack access to safely managed drinking water, and more than 4.2 billion lack safely managed sanitation (https://sdgs.un.org/topics/water-and-sanitation). With population expansion and climate change, this situation is likely to get worse as water shortages and floods become more common. How can green chemistry help?

On a global scale, the chemical sector is a major user of water, much of which is for heating and cooling, and is recycled.

EU Chemical Industry Organic Carbon Water Pollution

Total Organic Carbon (1000 metric tons):
- 2007: 24.5
- 2008: 22.2
- 2009: 16.9
- 2010: 19.6
- 2011: 20.2
- 2012: 15.6
- 2013: 16.3
- 2014: 16.9
- 2015: 16.8
- 2016: 19.1
- 2017: 17.8
- 2018*: 13.7
- 2019*: 12.4

Total organic carbon (as total C or COD/3)

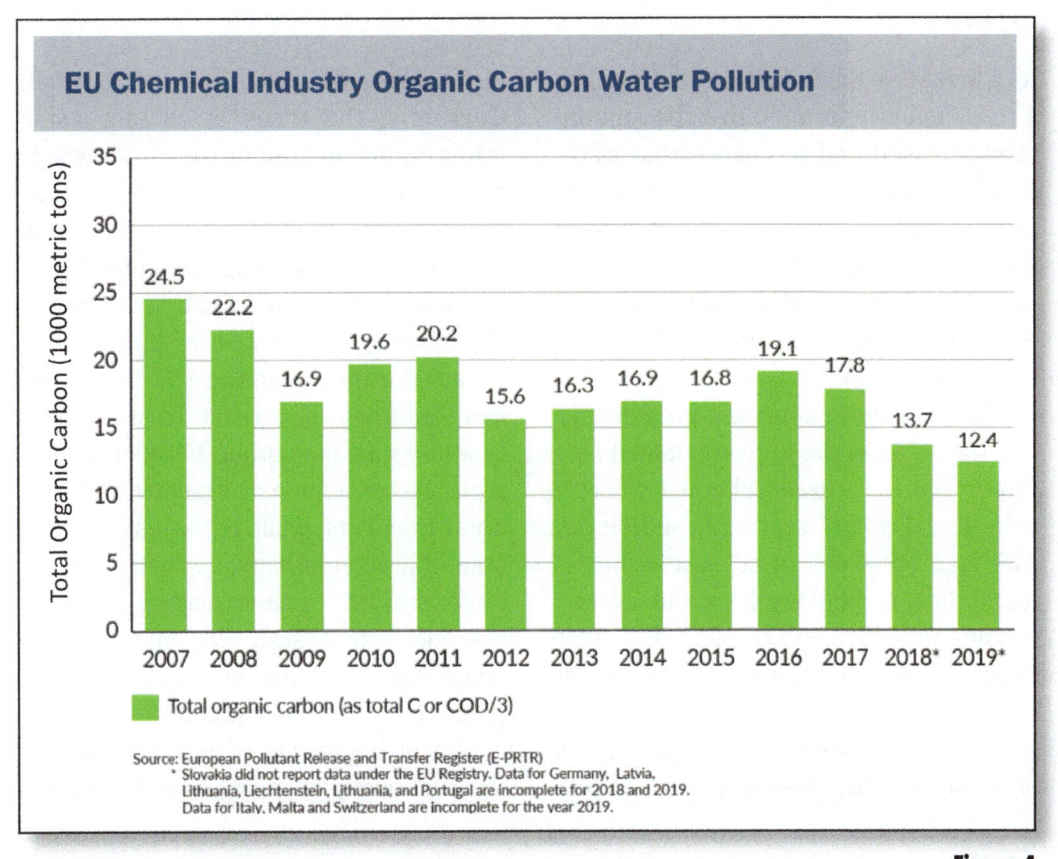

Source: European Pollutant Release and Transfer Register (E-PRTR)
* Slovakia did not report data under the EU Registry. Data for Germany, Latvia, Lithuania, Liechtenstein, Lithuania, and Portugal are incomplete for 2018 and 2019. Data for Italy, Malta and Switzerland are incomplete for the year 2019.

Figure 4

Role of Green Chemistry in Membrane Technology

Challenge	Role of Green Chemistry
Cost	New synthesis methods Development of new materials Analysis development to optimise pore size distribution
Fouling	Development of smooth, low charge surfaces Green Chemistry methods for producing effective • Coagulants • Biocides • Scale Inhibitors

Figure 5

As a first step, the industry therefore has a responsibility to minimize the amount of water used and to ensure that water returned to the environment is pollutant-free. In Europe, emission of water pollutants by the industry has almost halved since 2007, but work to minimize this and reduce water consumption must continue (**Figure 4**).

Large-scale desalination projects are one way to meet the growing demand for water, and here green chemistry has a role to play in improving efficiency. Although advances in reverse osmosis membrane technology have cut energy consumption by ~50% compared to older thermal processes, further improvement is needed to make the technology widely accessible, particularly in less-developed parts of the world (currently nearly half of the processes operated are thermal). Currently, the focus is on increasing water output and minimizing brine waste. Membrane technologies that enable high-pressure and high-salinity systems that are not prone to fouling are a priority.

A wide variety of membrane filtration technologies, including nanofiltration, are already starting to transform the water treatment industry (Najm and Trussell 1999). Low-pressure membrane filtration is now replacing conventional filtration for surface water treatment, while high-pressure membrane filtration is being evaluated for the removal of total organic matter. The main obstacle to large-scale implementation of membrane filtration is its capital cost (**Figure 5**). Ongoing innovations in the design of large-scale membrane systems are continually lowering capital cost and

> **Routes to Acrylamide**
>
> **Conventional**
>
> $C_3H_6 + NH_3 + 3/2O_2 = CH_2CHCN + H_2O$ Catalyst: bismuth phosphomolybdate; temperature 450 °C
>
> $CH_2CHCN + H_2O = CH_2CHC(O)NH_2$ Catalyst: copper salts, sulfuric acid; high temperature, high pressure, heavy metal waste
>
> **Green Route**
>
> $C_3H_8O_3 + NH_3 + O_2 = CH_2CHCN + H_2O$ Catalyst: vanadium pentoxide; temperature 400 °C
>
> $CH_2CHCN + H_2O = CH_2CHC(O)NH_2$ Catalyst: nitrile hydratase, high temperature, ambient pressure

Figure 6

making them increasingly cost competitive with conventional treatment processes. Advances in oxidation technology, including photocatalysis, are being developed to remove persistent pollutants.

Driven by rising population and stringent regulatory and sustainability mandates, the global water treatment chemicals market is estimated to double to over US$70 billion before the end of the decade (Verified Market Research 2021). It is therefore vital that existing and new products be manufactured and developed in line with green chemistry principles. One recent example is the use of biotechnology to produce acrylamide. Polyacrylamide (PAM) is one of the main flocculants used in sewage and industrial water treatment. This green process uses less energy, does not produce heavy metal waste. and is more cost effective (**Figure 6**). However, the PAM used is of high molecular weight, so there are possible pathways that release potentially carcinogenic acrylamide through polymer breakdown (Xiong et al. 2018). But new acrylamide-free flocculants are now being developed to overcome this potential issue.

Goal 12: Ensure sustainable consumption and production

Two issues the chemical industry has long faced are the amount of waste produced from processes and the fate of products from the industry at the end of their useful life. In the early days of green chemistry, Sheldon developed the concept of the *E-Factor*–the amount of waste produced per unit of product. He concluded that the further downstream the sector, the higher the E-Factor,

E-Factor: Waste Produced as a Proportion of Product

Industry Segment	Annual Production (te)	Kg By-products / Kg Products (E-Factor)	Approximate Total Waste (te)
Oil refining	10^6–10^8	Ca. 0.1	10^6
Bulk Chemicals	10^4–10^6	<1–5	10^5
Fine Chemicals	10^2–10^4	5–>50	10^4
Pharmaceuticals	10–10^3	25–>100	10^3

Figure 7   **Data from:** R.A. Sheldon. 1992. Chemistry & Industry (December 1): 904

Waste Produced by the Chemical Industry in the EU

- Production index
- Non-hazardous waste
- Hazardous waste

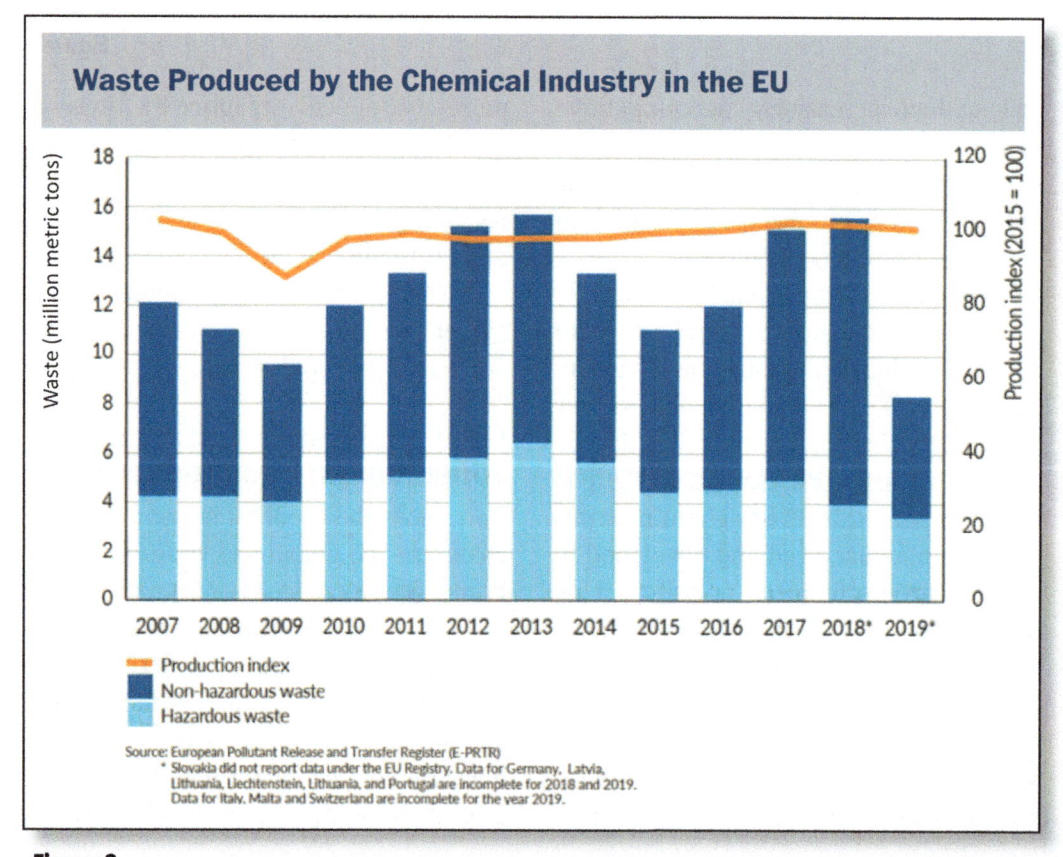

Source: European Pollutant Release and Transfer Register (E-PRTR)
* Slovakia did not report data under the EU Registry. Data for Germany, Latvia, Lithuania, Liechtenstein, Lithuania, and Portugal are incomplete for 2018 and 2019. Data for Italy, Malta and Switzerland are incomplete for the year 2019.

Figure 8

with fine chemicals having an E-Factor of up to 50 while bulk chemicals typically having one of less than 5 (**Figure 7**). This is not surprising, since it is in line with the complexity and number of process steps involved. Sheldon revisited his work 15 years later, in 2007, and concluded that not a lot had changed (Sheldon 2007, see https://www.sheldon.nl/roger/efactor.html). However, he did note that not all waste is the same and that it is much more important to eliminate hazardous waste than benign.

Waste production from the industry has been slow to reduce significantly—although, according to the latest figures from the European Chemical Industry Council (known by its French acronym, CEFIC) (**Figure 8**), 2019 was a good year.

With respect to managing products at the end of their useful life, recycling of polymers is one of the success stories of recent years, although the collection of household consumer products and contamination have been major issues. That said, recycling has generally been with lower-grade materials that are necessarily used in less-demanding applications, such as in the manufacture of garden furniture and fleece clothing. Green chemistry is now starting to solve this problem with the development of processes that, for example, are capable of manufacturing food-grade plastic bottles from enzymatically recycled polyethylene terephthalate (PET), such as procedures by Carbios (https://www.carbios.com/en/enzymatic-recycling/). Similarly, progress has been made into the so-called chemical recycling of plastic—plastic to monomers. Ideally, as oil refineries reduce their capacity in line with decreased needs for transportation fuel, we will be able to use these new recycled feedstocks in existing steam cracker equipment to shrink the carbon footprint further (**Figure 9**).

More progress is needed to close the

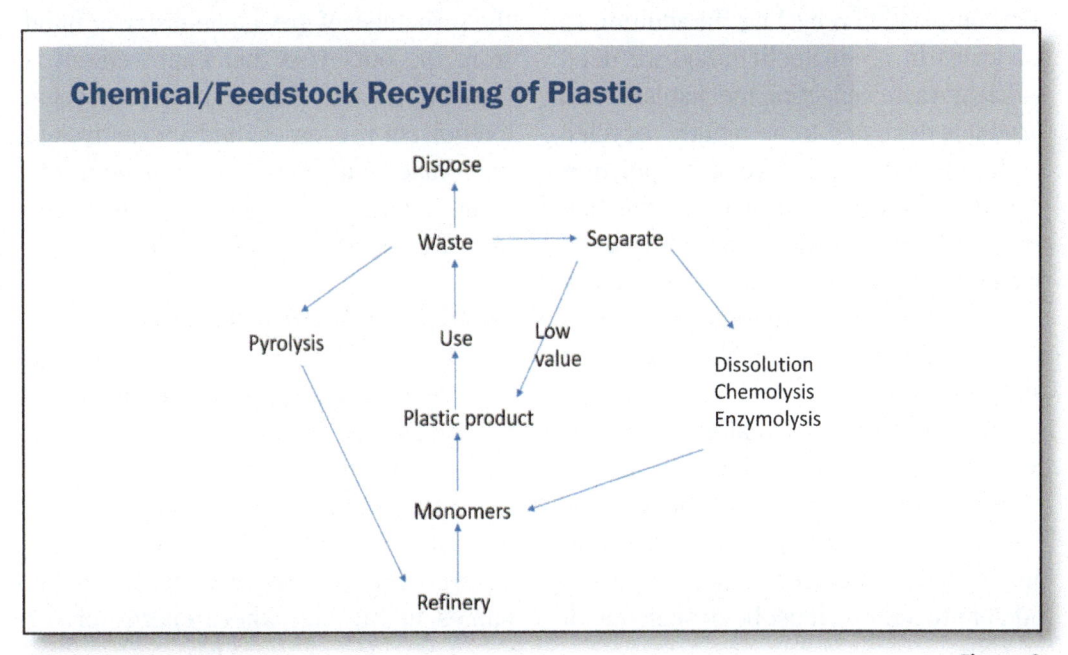

Plastic recycling schematic.

Figure 9

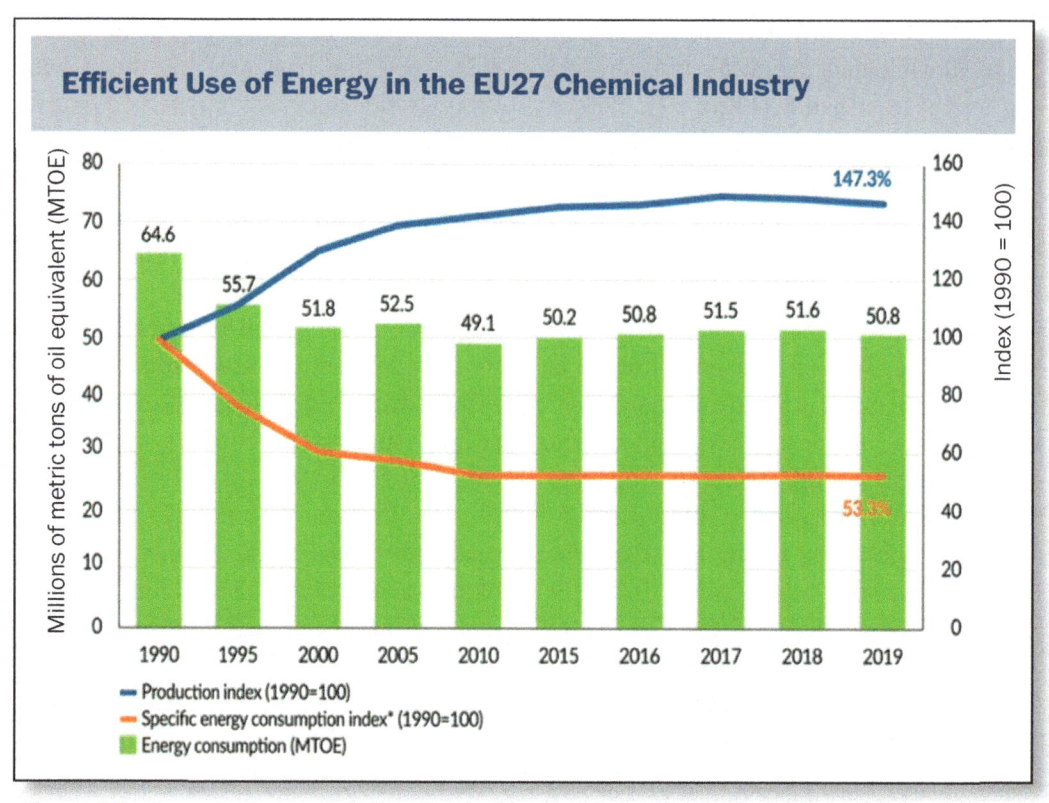

Efficient Use of Energy in the EU27 Chemical Industry

Figure 10 — EU27 specific energy consumption dropped by 47% during a 29-year period. (Source: CEFIC Facts and Figures Report 2022)

loop on plastics recycling. In addition to advances in green technology, we need efficient waste collection mechanisms and products designed to be readily recycled, both from the perspective of the polymer used and the construction of the finished article. Product design is probably one of the principles of green chemistry that has received the least attention, but it is a vital element of sustainable production and consumption.

As noted earlier, tackling the waste produced during chemical manufacture is proving problematic, and in many cases processes using existing technologies have already been optimized. The next generation of procedures must be designed with the principles of green chemistry in mind from the outset so that improvement is built in. In particular, we need to develop low-solvent processes, find alternatives to acid/base neutralizations that generate copious amounts of aqueous waste, and improve separation technologies.

Goal 13: Climate action

The commitment to achieving climate change targets has become the main focus of green chemistry for many companies and research groups.

Improving energy efficiency has been a key target for the chemical sector for several decades now, particularly for companies in Europe, where energy prices

Downstream Energy Savings from Using Products of the CI

- **Insulation** materials for the construction industry reduce heat loss by buildings and thus the use of heating fuel. Insulation alone accounted for 40% of the identified CO_2e savings

- Use of **chemical fertilizer and crop protection** in agriculture which increases crop yields, avoiding emissions from land use changes

- **Advanced lighting** solutions: LEDs and compact fluorescent lamps (CFLs), with longer lifetimes and greater luminous efficacy than incandescent bulbs, save significant energy

- The 7 next most important levers in 2005 were **plastic packaging, marine antifouling coatings, synthetic textiles, automotive plastics, low-temperature detergents, engine efficiency additives, and plastic piping**

Largest donwnstream emissions savings enabled by the chemical industry.   **Figure 11**

and associated green taxes are a major burden, and consumer demand for sustainable goods is high. Production in the European chemicals industry has grown by about 50% in the last 30 years, but there is now less energy used. Energy use per metric ton is now a little over 50% of 1990 levels (**Figure 10**).

This huge drop in specific energy consumption has been hard won. Much early progress was the result of companies tackling the easy, low-cost wins, such as reducing leaks, improving insulation, utilizing waste heat better, and installing more energy-efficient equipment, such as pumps and boilers (**Figure 11**). As time has passed, it has been necessary to look more at the chemistry itself.

For example, most titanium dioxide processes now operate using the more energy-efficient chloride method rather than the sulfate procedure. BP's development of the CATIVA process for acetic acid production has cut energy consumption by around 30% due to increased reactor productivity and reduction in energy-intensive distillation steps. The refinement of metallocene catalysts, first discovered almost 50 years ago, has led to significant cutbacks in the energy consumed during polyethylene manufacture, largely by increasing reactor throughput. Combining the latest catalyst technology with the use of bioethylene from bioethanol reduces

Li-ion Battery Schematic

e- charge ↔ e- discharge

Electrolyte

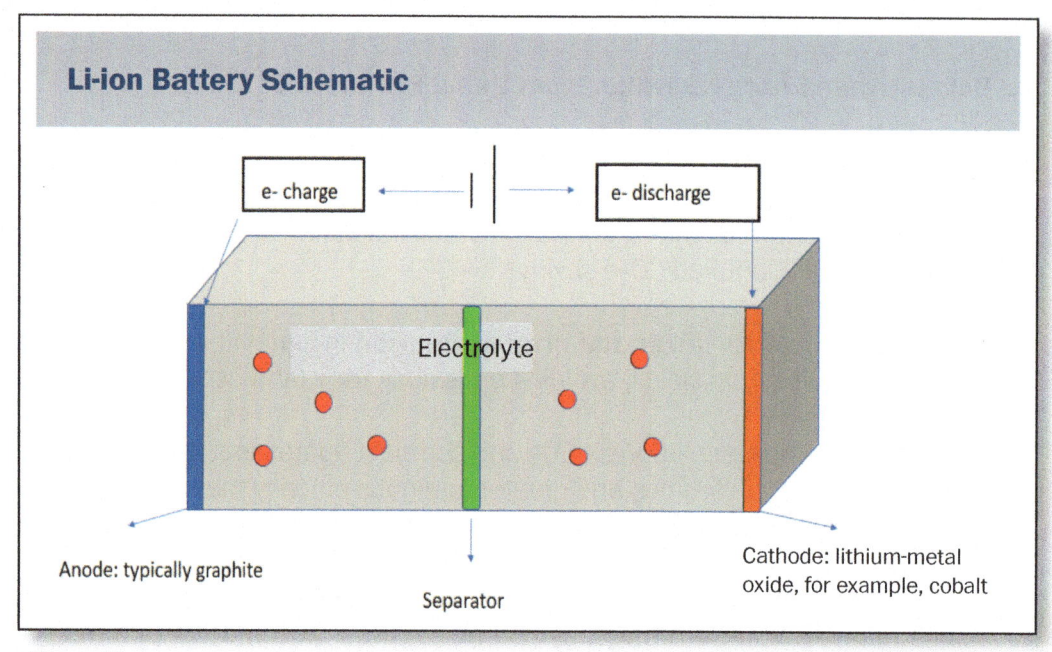

Anode: typically graphite

Separator

Cathode: lithium-metal oxide, for example, cobalt

Figure 12

the carbon footprint further. Examples like these of significant process improvements by making a step-change to energy consumption for bulk chemicals are relatively rare, but there are a multitude of examples where small changes to operating conditions, catalysts, solvents, and so on are making an important contribution.

The products of the chemical industry are also helping other sectors reduce their carbon footprint. A report by McKinsey commissioned by the International Council of Chemical Associations (2009) showed that for every metric ton of CO_2 emitted by the chemical industry, over 2 t of CO_2 was saved by the products and technologies provided to other industries or users.

Looking to the future, there are many other areas where green chemistry can have a significant impact, notably with respect to achieving net-zero carbon emissions.

Batteries: With the move to electric vehicles and the need for storage of renewable electricity from intermittent sources, there is a growing need for improved rechargeable battery technology. Chemistry has played a leading role in the development of more efficient, sustainable batteries from the early lead–acid technology through to the first dry cell zinc–carbon to modern lithium-ion (Li-ion) batteries (**Figure 12**). The main challenges for battery technologies are safety (lithium-ion batteries can ignite), improved power-to-weight ratio (to enable longer journeys without recharging), and better charging rates. Advances in safety are likely to be brought about by moving to solid-state batteries, in which lithium ions pass through a nonflammable polymer or inorganic electrolyte. These electrolytes may also be high voltage/high capacity, producing high-density, lighter batteries.

NanoBolt lithium tungsten batteries contain tungsten and carbon nanotubes that bond to the copper anode and build up a web-like nanostructure with a very large surface area. This enables more Li ions to attach to it during the recharge and discharge cycles, which makes recharging faster and also stores more energy (Aquion Energy). Currently, around 0.01% of the global supply of lithium is refined annually, which is expected to increase exponentially with subsequent high prices and shortage of refining capacity. Much work has gone into the development of sodium-ion technology, sodium being more naturally abundant and less expensive than lithium. However, it has a much lower energy density due to conventional carbon anodes absorbing lower amounts of sodium ions. A potential solution may be the use of graphene or nanocarbon anodes.

Whatever battery technology dominates in the future, there will be the need for recovery, processing, and recycling of the precious elements used to make the devices. Green chemistry will have a key role in making such processes efficient and commercially viable.

Hydrogen: With the necessary move away from fossil fuels, many countries are looking to hydrogen as a fuel for heating homes and buildings, running industrial boilers, and powering transportation, particularly for heavy public transport vehicles where battery technology is more difficult to apply. Natural gas reforming is a mature technology but only about 80% efficient in gas usage, limiting its appeal unless carbon capture and storage (CCS) and possibly reutilization of the released carbon dioxide can be deployed. From the green chemistry perspective, electrolysis of water offers the greatest medium-term promise for hydrogen production, provided the electricity comes from a renewable source. Although large-scale plants have been built, the cost of electricity currently prevents widespread commercialization of this technology, but many believe that renewable electricity will become less expensive over the medium term and may even be in surplus on occasion. Electrolysis of water has a significant kinetic barrier to overcome, and much work is focused on development of stable and efficient catalysts as well as on reducing the capital costs of electrolyzer units.

Other options being looked at are also dependent on green chemistry. They include the production of hydrogen from biomass and directly from sunlight using advanced photocatalysis. Concerning the use of hydrogen:

- The pace of adoption needs to increase to meet the 2050 net-zero targets.

- The cost gap between green-sourced hydrogen and hydrogen from fossil sources (a gap often differing by a factor of two to four, including CCS) must be reduced.

- The cost differential will reduce as renewables become more competitive.

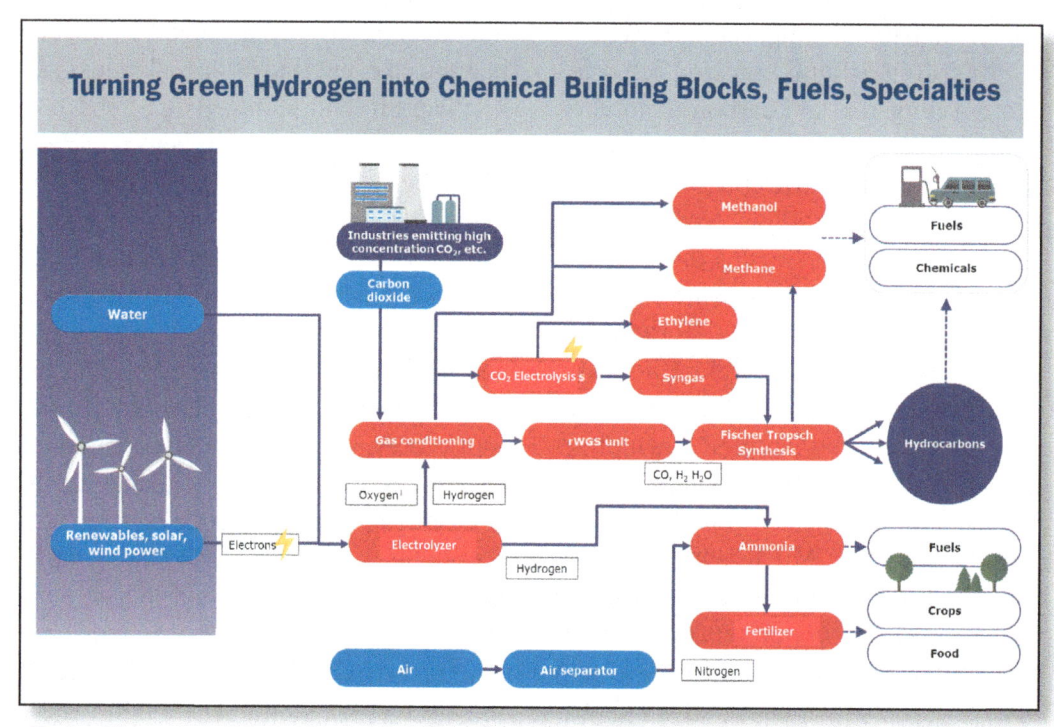

Figure 13   A chemical industry based on hydrogen and CO_2

- Strong collaborative innovation is needed to drive costs down.

- Very large-scale storage is potentially difficult.

- There is the potential for catalyst development and electrolyzer cost reduction.

- This technology may need to be combined with desalination technology.

- Biobased production is a possible alternative.

- There is the potential for direct production using sunlight—although advances in photocatalysis/semiconductors are needed.

CO_2 utilization: Growth in the use of hydrogen from natural gas will produce large volumes of CO_2. The majority of this will need to be stored, but there is a real opportunity to increase the use of CO_2 as a raw material for chemical production. Currently, urea is the main organic chemical produced from CO_2, while some poly(propylene) carbonate is also manufactured. Production of inorganic carbonates is a likely significant future use of CO_2 and will "lock" the carbon away in cement, for example.

Coming perhaps soon are commercial-scale plants for other products, such as methanol. In fact, there is already a small-scale facility in Iceland, and the first commercial green methanol plant based on captured CO_2 and green hydrogen was recently announced

Biorefinery Concept

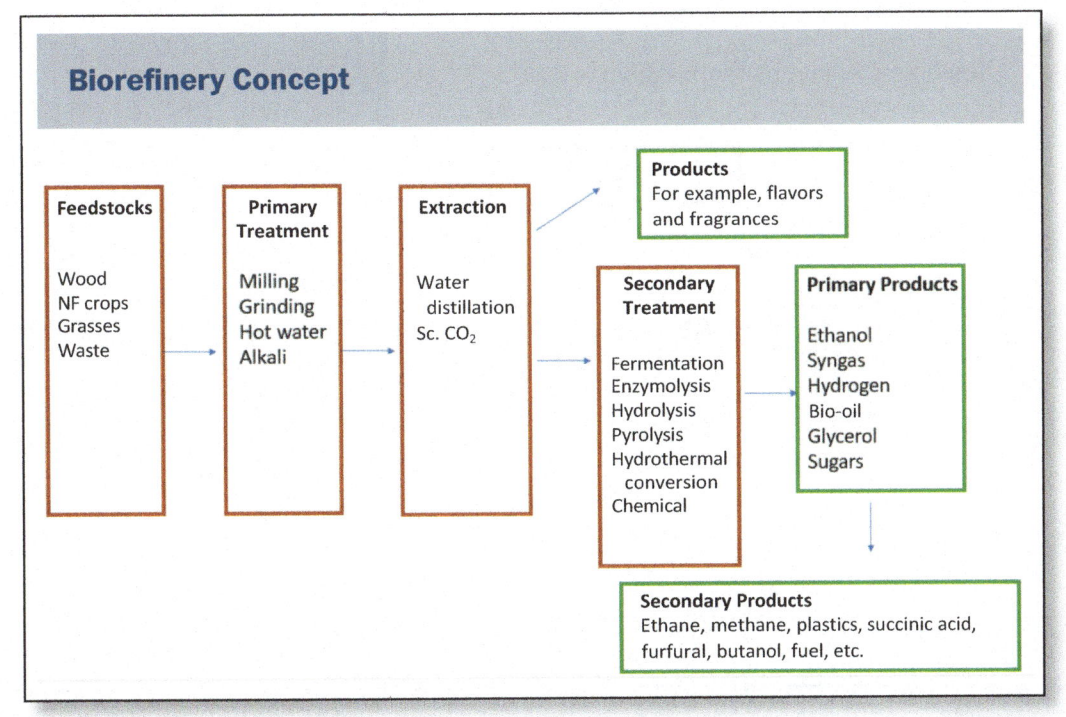

Feedstocks	Primary Treatment	Extraction		
Wood NF crops Grasses Waste	Milling Grinding Hot water Alkali	Water distillation Sc. CO_2		

Products: For example, flavors and fragrances

Secondary Treatment: Fermentation, Enzymolysis, Hydrolysis, Pyrolysis, Hydrothermal conversion, Chemical

Primary Products: Ethanol, Syngas, Hydrogen, Bio-oil, Glycerol, Sugars

Secondary Products: Ethane, methane, plastics, succinic acid, furfural, butanol, fuel, etc.

NF = non-food; **Sc** = supercritical

Figure 14

(https://matthey.com/en/news/2021/jm-technology-selected-for-worlds-first-climate-neutral-methanol-plant). Yet more commercial products that are likely to be developed in the near future using CO_2 include formic acid, organic carbonates and carbamates, and other organic acids and alcohols. Indeed, pathways to a new green petrochemicals industry based on methanol from hydrogen and captured CO_2 have already been mapped out (**Figure 13**).

Biobased chemicals: As noted above, many products of the chemical industry help other sectors reduce their carbon footprint. By using biobased materials (nonfood crops and waste), the carbon footprint of both the producer and end user will be reduced further. There has been a huge growth in research in this area, with some notable commercial developments, not least in biofuels and polymers such as polylactates, sorbitol, and biosuccinic acid. The biorefinery concept (**Figure 14**) has been around for over 20 years, but there are relatively few examples of large-scale, multiproduct biorefineries of sizes similar to those of oil refineries, with most being bioethanol or biodiesel factories. Oil prices that were relatively low have undoubtedly hindered development in this area, but the postpandemic hike in oil and gas prices should renew interest. Changeable government policies, the continuing debate over land use, and the dominant need for biofuels are also not conducive to large-scale investment in multiproduct refineries.

We can take lignin as an example—a

Possible Chemicals from Lignin

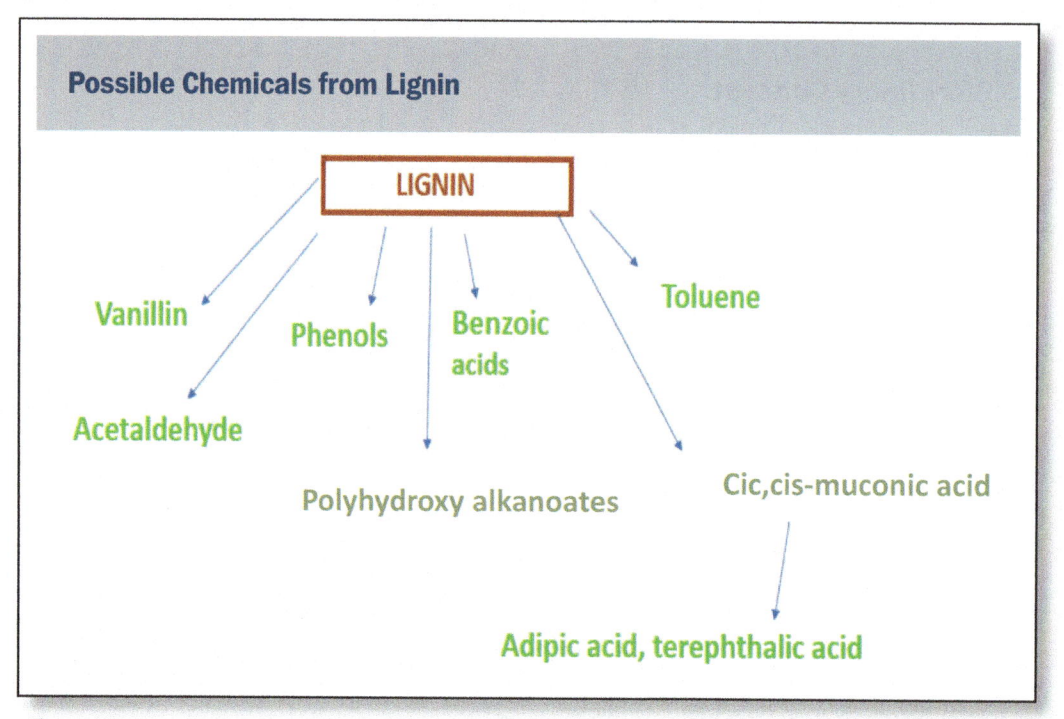

LIGNIN → Vanillin, Phenols, Benzoic acids, Toluene, Acetaldehyde, Polyhydroxy alkanoates, Cic,cis-muconic acid, Adipic acid, terephthalic acid

Figure 15

multimillion–metric ton waste product from the pulp and paper industry. The use of lignin for chemical production has been limited (it is used in the manufacture of vanillin) due to contamination from salts, carbohydrates, particulates, volatiles, and the wide variability of molecular weight distribution of lignosulfonates. However, lignin has huge potential in the production of aromatic hydrocarbons and phenols (Paone, Tabanelli, and Mauriello 2020) (**Figure 15**). But to make a lignin biorefinery more commercially viable, more robust depolymerization catalysts are needed, as well as a change in commercial drivers.

Conclusion

Progress in green chemistry has underpinned our gradual evolution to more sustainable practice in many areas, and there are signs that both knowledge and implementation of green chemistry practices are growing more rapidly. Globally, there has been a huge increase in relevant education, but we are still some way off from green chemistry becoming synonymous with chemistry—the complete integration of green chemistry in education and industrial practice.

Industrially, there are many examples of processes being more energy and resource efficient, and the industry is now safer than it has ever been. Regulation, stakeholder demands, and a growing realization that green chemistry and sustainability make commercial sense are driving this trend, particularly in the areas of waste reduction and energy efficiency.

Since the concept of green chemistry was first introduced, climate change has become widely recognized as the largest medium-term threat to society. By adopting the green chemistry approach, the chemical industry can make a significant contribution to reducing the carbon footprint of other sectors while at the same time pursuing its own path to net zero. Important future developments include much more efficient battery technology, advances in electrolytic hydrogen production, and significant growth in chemical production from biomass, CO_2, and H_2. Green chemistry has a role to play in solving many global issues identified in the UN Sustainable Development Goals, not the least of which are the growing challenges of sustainable production and consumption and sustainable management of water caused by a growing population. Going forward, it can help us to square the circle of maintaining living standards while reducing our impact on the planet.

Now more than ever, green chemistry is needed to help solve the major issues facing society as the effects of climate change become more apparent, the global population continues to grow, and living standards and consumerism continue to rise. ◊

References

Anastas, P.T., and J.C. Warner. 1998. *Green Chemistry: Theory and Practice.* Oxford, United Kingdom: Oxford University Press.

Aquion Energy. 2022. "Battery Innovations and Technology Powering Our Future." *Aquion Energy.* Accessed on September 9, 2022. https://www.aquionenergy.com/technology/.

European Chemical Industry Council. 2022. "The European Chemical Industry—A Vital Part of Europe's Future: Facts & Figures 2022." Accessed on September 9, 2022. https://cefic.org/app/uploads/2022/01/Leaflet-FactsFigures_interactif_V02.pdf.

International Council of Chemical Associations. 2009. "Innovations for Greenhouse Gas Reductions: A Life Cycle Quantification of Carbon Abatement Solutions Enabled by the Chemical Industry." Accessed on September 9, 2022. https://web.archive.org/web/20220618121045/https://www.nikkakyo.org/sites/default/files/icca_LCA_report2009_en.pdf

Najm, I., and R.R. Trussell. 1999. "New and Emerging Drinking Water Treatment Technologies." In *Identifying Future Drinking Water Contaminants*, edited by the National Research Council, 220–243. Washington, DC: The National Academies Press.

Paone, E., T. Tabanelli, and F. Mauriello. 2020. "The Rise of Lignin Biorefinery." *Current Opinion in Green and Sustainable Chemistry* 24 (August): 1–6. Accessed on October 1, 2022. https://doi.org/10.1016/j.cogsc.2019.11.004.

Sheldon, R.A. 1992. "Organic Synthesis: Past, Present, and Future." *Chemistry & Industry* 19: 903–906.

———. 2007. "The E Factor: Fifteen Years On." *Green Chemistry* 9: 1273–1283. Accessed on October 1, 2022. https://doi.org/10.1039/B713736M.

Verified Market Research. December 2, 2021. "Water Treatment Chemicals Market Size Worth $74.92 Billion, Globally, by 2028 at 8.5% CAGR: Verified Market Research®." PR Newswire. Accessed on September 9, 2022. https://www.prnewswire.co.uk/news-releases/water-treatment-chemicals-market-size-worth-74-92-billion-globally-by-2028-at-8-5-cagr-verified-market-research-r–821996219.html.

Xiong, B., R.D. Loss, D. Shields, T. Pawlik, R. Hochrreiter, A.L. Zydney, and M. Kumar. 2018. "Polyacrylamide Degradation and Its Implications in Environmental Systems." *npj Clean Water* 1: 17. Accessed on October 1, 2022. https://doi.org/10.1038/s41545-018-0016-8.

Session 2 | First Commentary
on the Presentation by the Hon. Mike Lancaster

Green Chemistry: Challenges and Opportunities

Professor James Clark
Chemistry, University of York,
Director, Green Chemistry Centre of Excellence and
Circa Renewable Chemistry Institute, UK

This paper looks at the current problems facing the chemical and allied industries and how we can use renewable resources and waste valorization to solve many of these problems.

We live in a world of chemicals and of the chemistry that produces those chemicals to create all of the articles of modern society. The demand for chemicals continues to increase and parallels a growth in the value and volume of chemicals, with millions of new chemicals becoming available every year (Kümmerer, Clark, and Zuin 2020). This not only leads to more products but to products with more and more ingredients. A simple shower gel illustrates this, with some consumer products containing more than 50 different chemicals. This is not sustainable—in products, processes, or resources.

We use largely nonrenewable resources in mostly wasteful and hazardous processes to make products that are often toxic and environmentally harmful. The low-carbon revolution that is being driven by escalating fears over climate change further increases this demand and requires more new chemicals—and especially metals little used previously (Lotzof 2020). A modern mobile phone can contain ele-

About the Speaker

- Professor of Chemistry; Founding Director, Green Chemistry Centre of Excellence; Director, Circa Renewable Chemistry Institute, University of York, UK.
- Chair Professor, Fudan University, China.
- Director and CEO, JHC Sustainability Ltd., UK.
- Recipient of the 2018 Royal Society of Chemistry Green Chemistry Prize and 2021 European Sustainable Chemistry Award.
- Scientific Advisor, Circa Group, Norway.
- PhD in Chemistry, King's College London, UK.

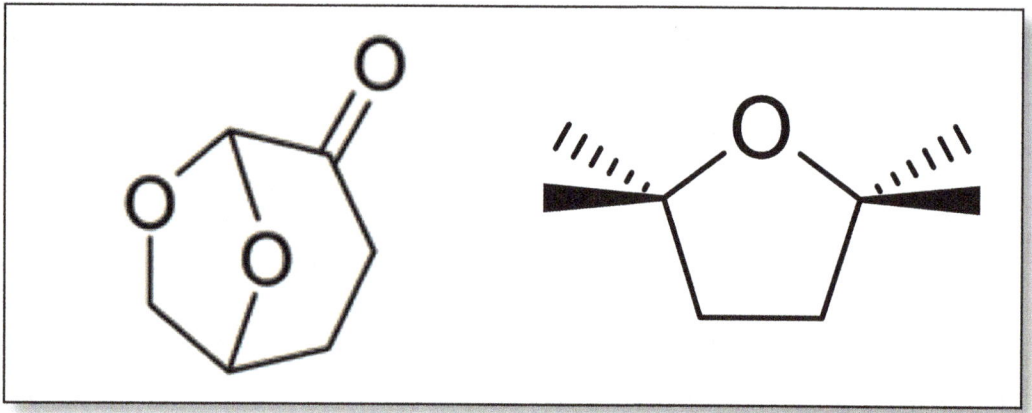

Figure 1. Two new nontoxic, "bio better" commercial solvents: dihydrolevoglucosenone (Cyrene) (left) and 2,2,5,5-tetramethyloxolane (TMO)

ments from close to half of the periodic table.

Our global reliance on metals is growing. Working toward carbon neutrality is leading to our being metals dependent, yet we recycle very little. This concern is amplified by current global crises and makes us reconsider the complex and lengthy global supply chains that feed our industries. Consequently, there is a growing call for more local resources and more regional supply chains, and since more and more of those resources are being wasted, we have an opportunity to valorize those wastes and feed industry and consumer demand through a circular and more localized economy (Morone and Clark 2020).

Biobased chemicals

Over 90% of the organic chemicals we use come from petroleum, which has been the dominant source of carbon for industry for the last hundred years. However, petroleum is nonrenewable, its supply is concentrated in politically problematic regions, and its exploitation to make chemicals is based largely on old, inefficient, and hazardous chemistry. Biomass is a renewable source of carbon, available across the globe, and with the built-in chemical functionality that can negate the need for some of the worst chemical processes (Calvo-Serrano et al. 2019). The value of biomaterials has recently been highlighted by McKinsey & Company, which sees a large and growing market for biobased chemicals (Brennan, Chui, and Wen 2021).

While drop-ins and bioreplacements can help wean industry off petrochemicals, we should aim for new, biobased chemicals that are "bio better," offering green credentials and improved performance. Good examples of this are the new solvents Cyrene (dihydrolevoglucosenone) from Circa (https://circa-group.com/products/cyrene/) and TMO (2,2,5,5-tetramethyloxolane) from Addible (https://www.addible.co.uk/), which were designed to replace traditional, toxic petrosolvents (**Figure 1**). These new solvents can be made from biomass and can replace solvents like NMP (N-methyl-2-pyrrolidone), DMF (dimethyl-

The uptake of metals from waste using biomass followed by conversion to a chemical catalyst **Figure 2**

formamide), and toluene in many applications, and also can offer improved performance in some applications.

Metals from waste

Like oil, virgin metals are coming at increasing economic and environmental costs. Metal deposits are becoming lower grade and harder to reach. This can mean increasing downstream processing, adding to the economic costs but also consuming more resources and creating more waste. The answer could be in the waste—most of our consumer goods end up as waste, with little useful recycling. Most of the metals that we use are hardly recycled at all. We need to see these metal-rich wastes as sources of future metals—and there are technologies available to mine these. Successful examples include the use of plants that can be natural metal scavengers, and the utilization of porous solids that can capture and retain metals from waste streams (**Figure 2**).

Conclusions

We can tackle the embedded problems of traditional manufacturing—nonrenewable resources, inefficient manufacturing, increasing waste, and pollution—through a more intelligent use of resources as part of a circular economy. Chemicals and chemistry are a vital part of this. If we ensure that our progress is genuinely green through good metrics, including giving credit for recycling (Lokesh et al. 2020), then we can look forward to a sustainable future. ◊

References

Brennan, T., M. Chui, W. Chyan, and A. Spamann. November 18, 2021. "The Third Wave of Biomaterials: When Innovation Meets Demand." McKinsey & Company. Accessed on September 8, 2022. https://www.mckinsey.com/industries/chemicals/our-insights/the-third-wave-of-biomaterials-when-innovation-meets-demand.

Calvo-Serrano, R., M. Guo, C. Pozo, Á. Galán-Martín, and G. Guillén-Gosálbez. 2019. "Biomass Conversion into Fuels, Chemicals, or Electricity? A Network-Based Life Cycle Optimization Approach Applied to the European Union." *ACS Sustainable Chemistry & Engineering* 7 (12): 10,570–10,582. https://doi.org/10.1021/acssuschemeng.9b01115.

Kümmerer, K., J.H. Clark, and V.G. Zuin. 2020. "Rethinking Chemistry for a Circular Economy." *Science* 367 (6476): 369–370. https://doi.org/10.1126/science.aba4979.

Lokesh, K., A.S. Matharu, I.K. Kookos, D. Ladakis, A. Koutinas, P. Morone, and J. Clark. 2020. "Hybridised Sustainability Metrics for Use in Life Cycle Assessment of Bio-Based Products: Resource Efficiency and Circularity." *Green Chemistry* 22 (3): 803–813. http://www.doi.org/10.1039/C9GC02992C.

Lotzof, K. October 7, 2020. "Your Mobile Phone Is Powered by Precious Metals and Minerals." National History Museum. Accessed September 9, 2022. https://www.nhm.ac.uk/discover/your-mobile-phone-is-powered-by-precious-metals-and-minerals.html.

Morone, P., and J.H. Clark, eds. 2020. *Transition Towards a Sustainable Biobased Economy.* Green Chemistry Series. Croydon, UK: The Royal Society of Chemistry.

Session 2 | Second Commentary
on the Presentation by the Hon. Mike Lancaster

Renewable Phenols as Feedstock for Polymers – India's Potential

Professor Bimlesh Lochab
Department of Chemistry, School of Natural Sciences
Shiv Nadar University, Delhi-NCR, India

Hello, I am Dr. Bimlesh Lochab from Shiv Nadar University. It is a pleasure to provide commentary on Hon. Lancaster's talk about the promise of green chemistry. I am going to specifically address the possible chemicals from lignin, and I will focus more on the phenolic compounds, because the phenols have a wide arena of applications, ranging from energy to materials to biological uses.

In my research group, we employ the tenets of green chemistry. As indicated in **Figure 1**, renewable resources came to the fore long ago. Indeed, renewable feedstocks were on the rise back in the 1850s. But then the coal revolution erupted, which suppressed the demand for renewables. This was followed by industrial economies moving toward a better alternative—natural gas and oil. But as we all know, these are finite, nonrenewable resources, so we have to look for superior alternatives. In this direction, we have to close the loop, and we discover that greener solutions are the best.

In my research group, we are looking for polymers derived from renewable phenols and are exploring a variety of applications. In this direction, we are looking toward polybenzoxazines, which are far

About the Speaker

- Professor and Head, Department of Chemistry, Shiv Nadar University, India.
- Researches the synthesis, functionalization, and conjunction of nano- and polymeric materials using the tenets of green chemistry.
- Fellow of the Royal Society of Chemistry (FRSC).
- Member, Subject Expert Committee on Chemical Sciences, Women Scientists Scheme-A, Department of Science and Technology, India.
- Member, National Advisory Committee, Asian Polymer Association.
- DPhil in Organic Chemistry, University of Oxford, UK.

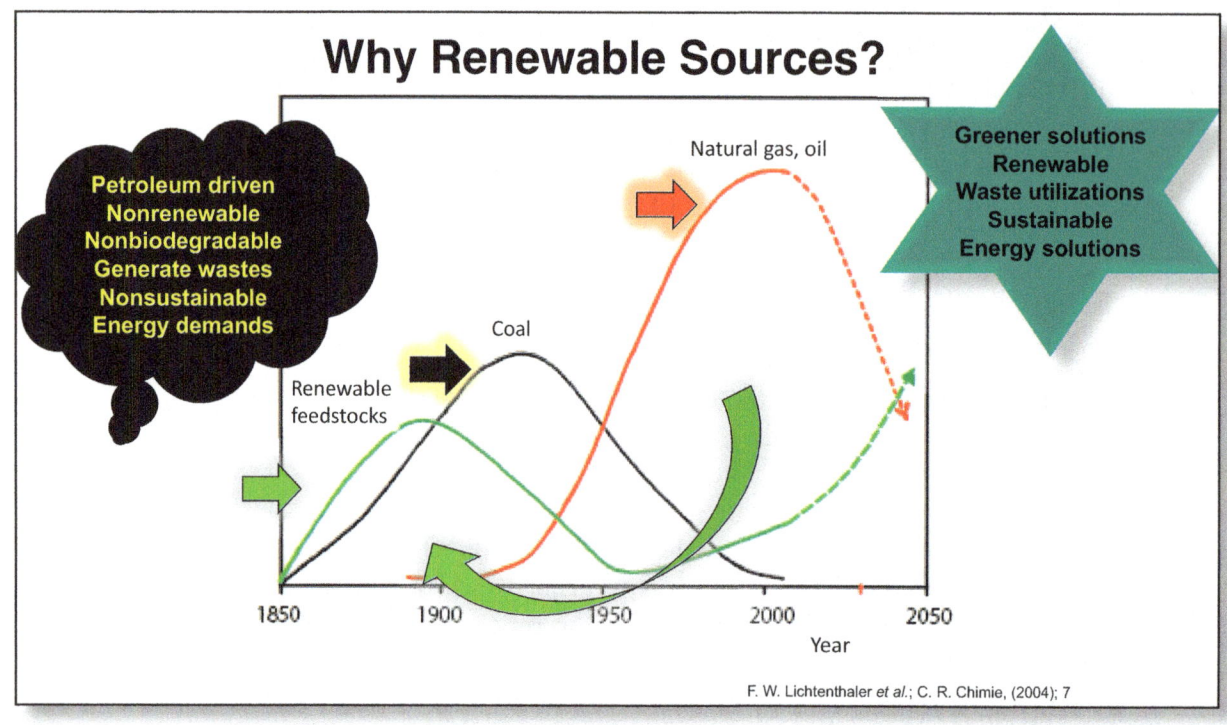

Figure 1 Adapted from Lichtenthaler, F.W., and S. Peters. 2004. "Carbohydrates as Green Raw Materials for the Chemical Industry." *Comptes Rendus Chimie* 7 (2): 65–90. https://doi.org/10.1016/j.crci.2004.02.002.

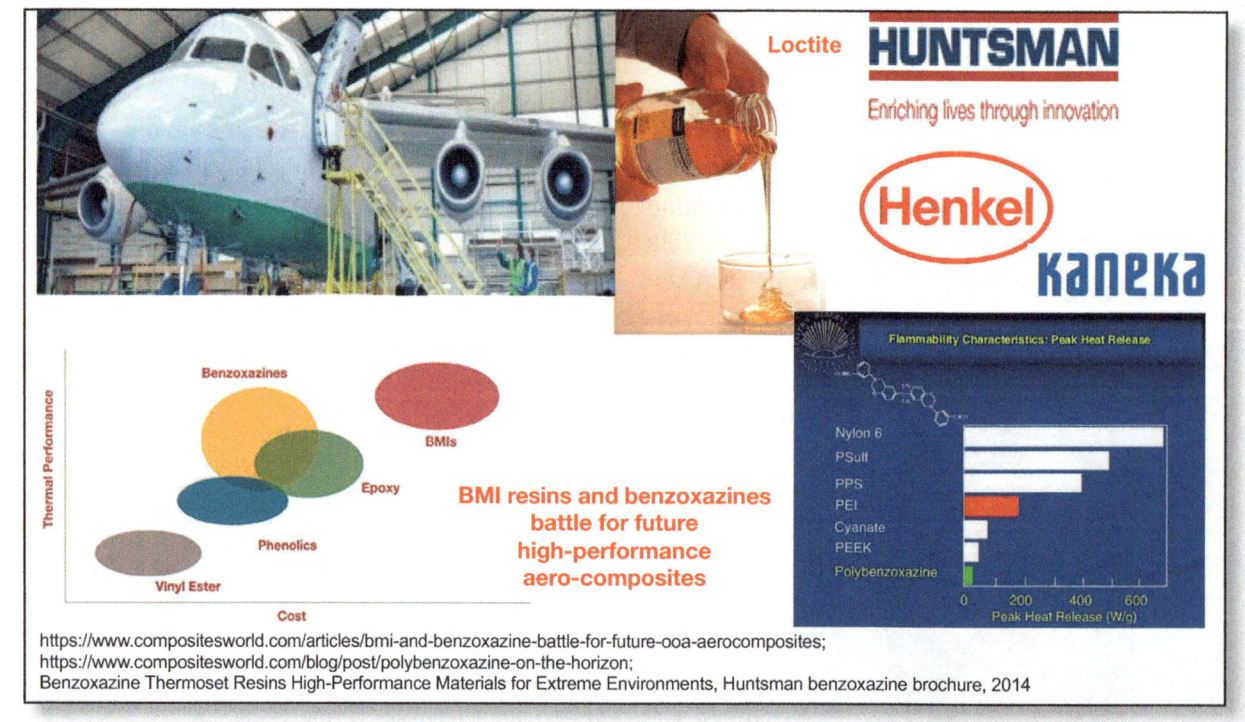

https://www.compositesworld.com/articles/bmi-and-benzoxazine-battle-for-future-ooa-aerocomposites;
https://www.compositesworld.com/blog/post/polybenzoxazine-on-the-horizon;
Benzoxazine Thermoset Resins High-Performance Materials for Extreme Environments, Huntsman benzoxazine brochure, 2014

Figure 2

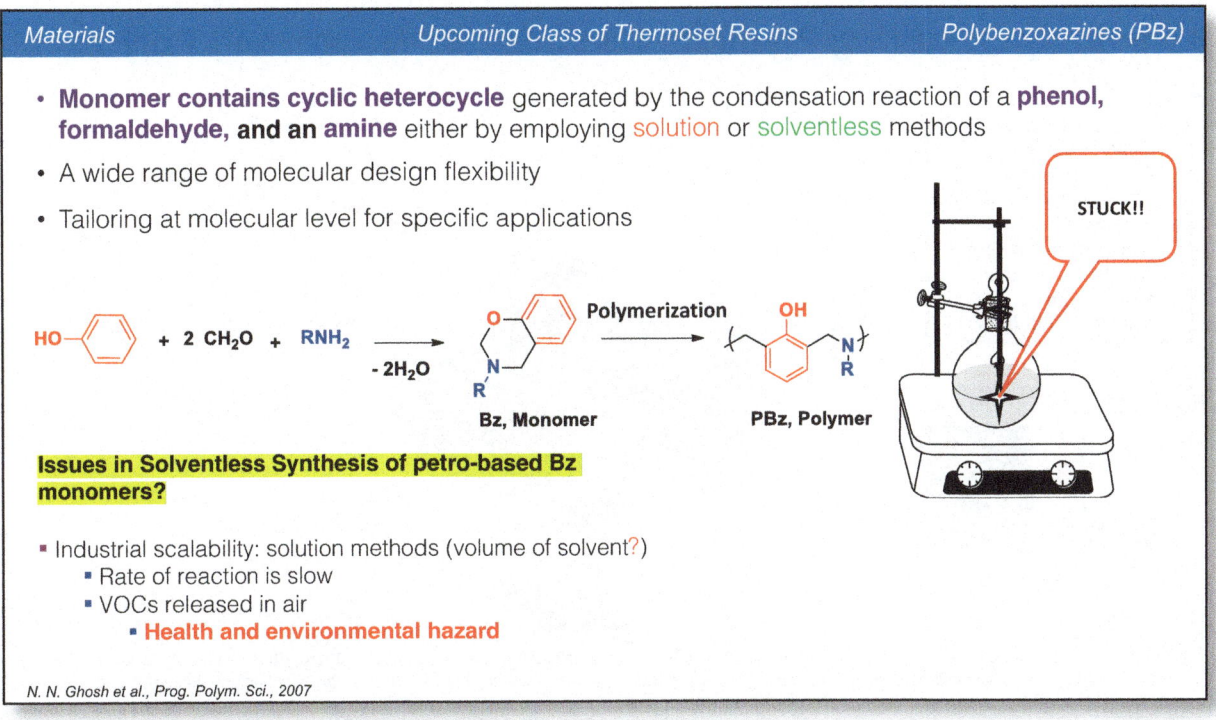

Professor Bimlesh Lochab

Materials — Upcoming Class of Thermoset Resins — **Polybenzoxazines (PBz)**

- **Monomer contains cyclic heterocycle** generated by the condensation reaction of a **phenol, formaldehyde, and an amine** either by employing solution or solventless methods
- A wide range of molecular design flexibility
- Tailoring at molecular level for specific applications

HO–⟨phenol⟩ + 2 CH$_2$O + RNH$_2$ —(–2H$_2$O)→ Bz, Monomer —Polymerization→ PBz, Polymer

STUCK!!

Issues in Solventless Synthesis of petro-based Bz monomers?

- Industrial scalability: solution methods (volume of solvent?)
  - Rate of reaction is slow
  - VOCs released in air
    - Health and environmental hazard

N. N. Ghosh et al., Prog. Polym. Sci., 2007

Bz: benzoxazine; **PBz**: polybenzoxazine; **VOC**: volatile organic compound

Figure 3

---

better in performance and cost than the traditional phenolic polymers (**Figure 2**). These phenol-based polybenzoxazines are our latest entry in the field of aerocomposites, and they are offering very good properties in terms of flammability characteristics.

What are polybenzoxazines? They contain an oxazine ring, which is obtained by a condensation reaction of a phenol, a formalin, and an amine molecule. We vary this phenol component or the amine component, and recently we also started substituting the formalin. From these reactions, we get a tailored monomer structure according to the particular need, and we start achieving a variety of properties. Then these compounds undergo ring opening polymerization and form the polymer (**Figure 3**). But while we were working with petroleum-based phenols and were performing these reactions under solventless conditions, the reaction stirrer became stuck after a while. We found that solventless synthesis of petroleum-based benzoxazine monomers is one of the major culprits.

In this direction, when we are looking for industrial scalability, solution methods are far better. However, while we are considering industrial-level synthesis of these monomers, the volume of solvent is a big question. When we are dealing with that, then the rate of reaction is slow and volatile organic compounds are released, creating a big issue from a health perspective.

Thus, we started exploring nonpetroleum-based, naturally occurring phenolic

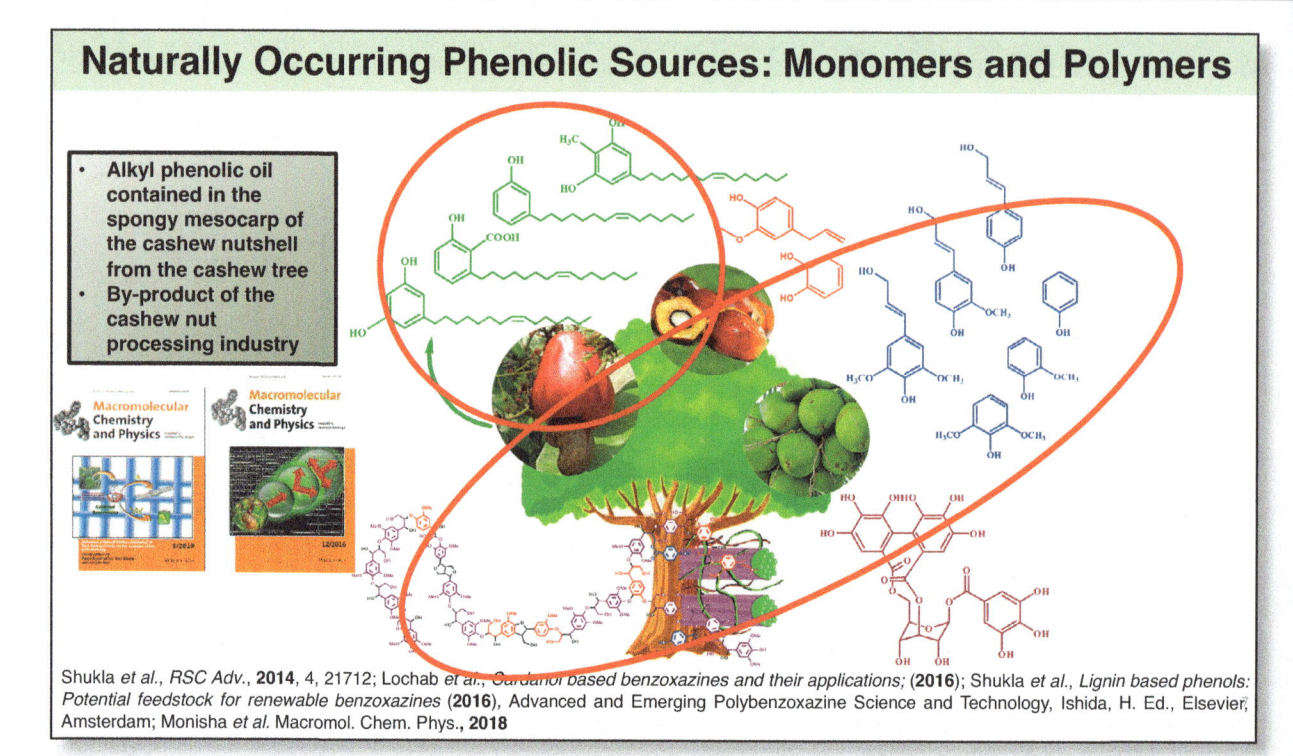

118   Second Commentary on the Presentation by the Hon. Mike Lancaster

Figure 4

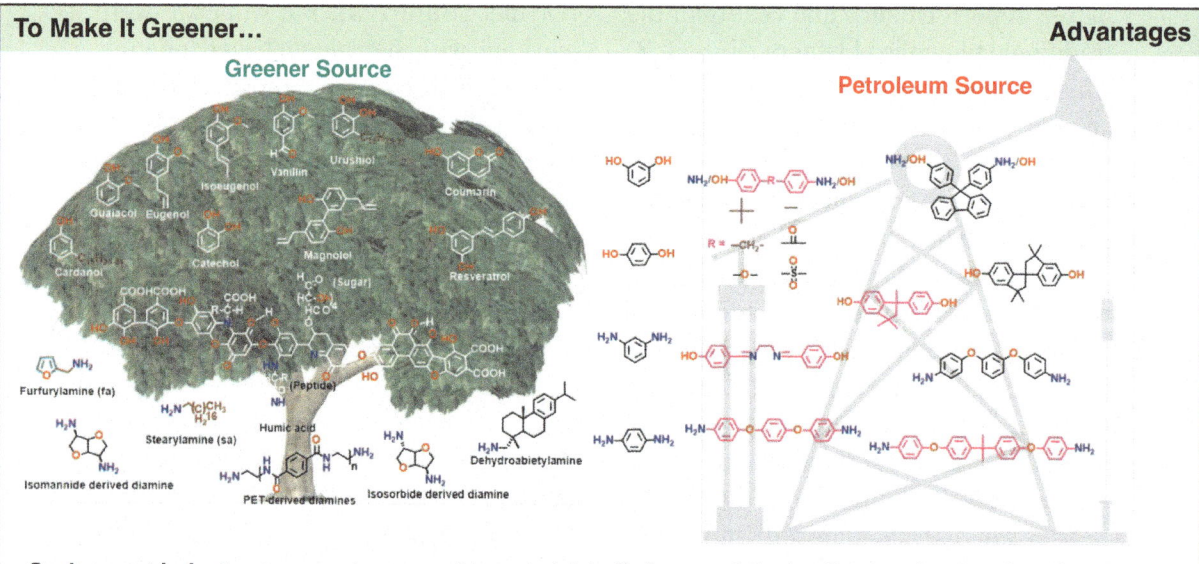

Figure 5

Applications — Adhesion

Higher LSS = better adhesive

- Mono-benzoxazine
- Bis-benzoxazine
- Tris-benzoxazine
- PF resin

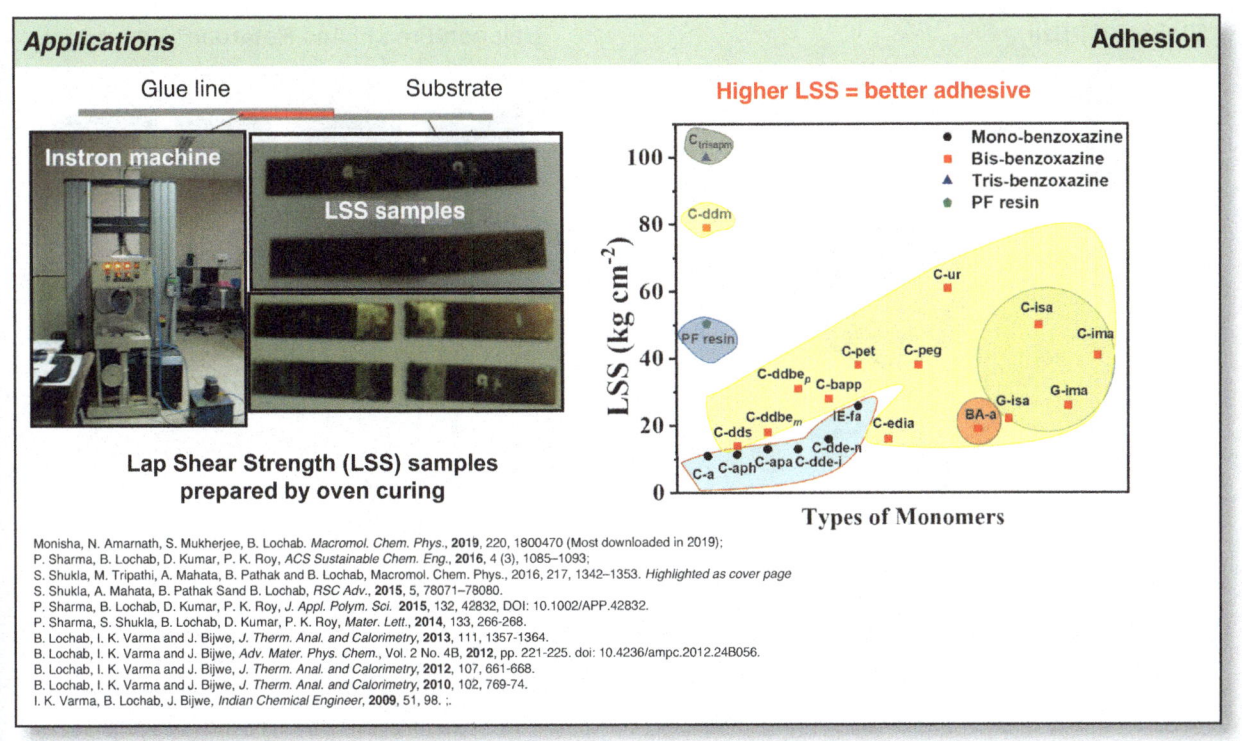

Lap Shear Strength (LSS) samples prepared by oven curing

Monisha, N. Amarnath, S. Mukherjee, B. Lochab. *Macromol. Chem. Phys.*, **2019**, 220, 1800470 (Most downloaded in 2019);
P. Sharma, B. Lochab, D. Kumar, P. K. Roy, *ACS Sustainable Chem. Eng.*, **2016**, 4 (3), 1085–1093;
S. Shukla, M. Tripathi, A. Mahata, B. Pathak and B. Lochab, Macromol. Chem. Phys., 2016, 217, 1342–1353. *Highlighted as cover page*
S. Shukla, A. Mahata, B. Pathak Sand B. Lochab, *RSC Adv.*, **2015**, 5, 78071–78080.
P. Sharma, B. Lochab, D. Kumar, P. K. Roy, *J. Appl. Polym. Sci.* **2015**, 132, 42832, DOI: 10.1002/APP.42832.
P. Sharma, S. Shukla, B. Lochab, D. Kumar, P. K. Roy, *Mater. Lett.*, **2014**, 133, 266-268.
B. Lochab, I. K. Varma and J. Bijwe, *J. Therm. Anal. and Calorimetry*, **2013**, 111, 1357-1364.
B. Lochab, I. K. Varma and J. Bijwe, *Adv. Mater. Phys. Chem.*, Vol. 2 No. 4B, **2012**, pp. 221-225. doi: 10.4236/ampc.2012.24B056.
B. Lochab, I. K. Varma and J. Bijwe, *J. Therm. Anal. and Calorimetry*, **2012**, 107, 661-668.
B. Lochab, I. K. Varma and J. Bijwe, *J. Therm. Anal. and Calorimetry*, **2010**, 102, 769-74.
I. K. Varma, B. Lochab, J. Bijwe, *Indian Chemical Engineer*, **2009**, 51, 98.

Figure 6

compounds—and India is rich in these, especially cashew nut fruit. **Figure 4** depicts cardanol, a liquid derived from cashew nutshells that has a long alkylene chain and a phenolic group. It can be a good substitute for phenolic polymers—and so can lignin and other corresponding phenolic compounds sourced from lignin.

We are also exploring other phenolic sources (**Figure 5**) to make our chemistry greener. This is proving to be a natural progression, as India is largely an agrarian country. Indeed, India is one of the world's leading producers of cardanol, which is available at a considerably lower cost than petroleum-based phenols—and it is nontoxic in nature.

As I mentioned, cardanols are characterized by a long alkylene chain. They can be used to pursue reactions in solventless conditions—and simultaneously this petroleum-sourced chemistry can be overwritten. We are using these monomers in adhesive applications, and are investigating whether we can replace traditional phenol formaldehyde resins with this new class of polybenzoxazines sourced from cardanol. We started in this direction, as you can see in **Figure 6**, with a bisphenol A–based resin formed from petroleum-sourced compounds. We found that the cardanol-based resin, created by greatly advantageous solventless processing, offers a much higher Lap Shear Strength value, which is a measure of adhesive strength that we determined using an Instron machine.

The synthesis of the benzoxazine monomer, its polymerization, and the

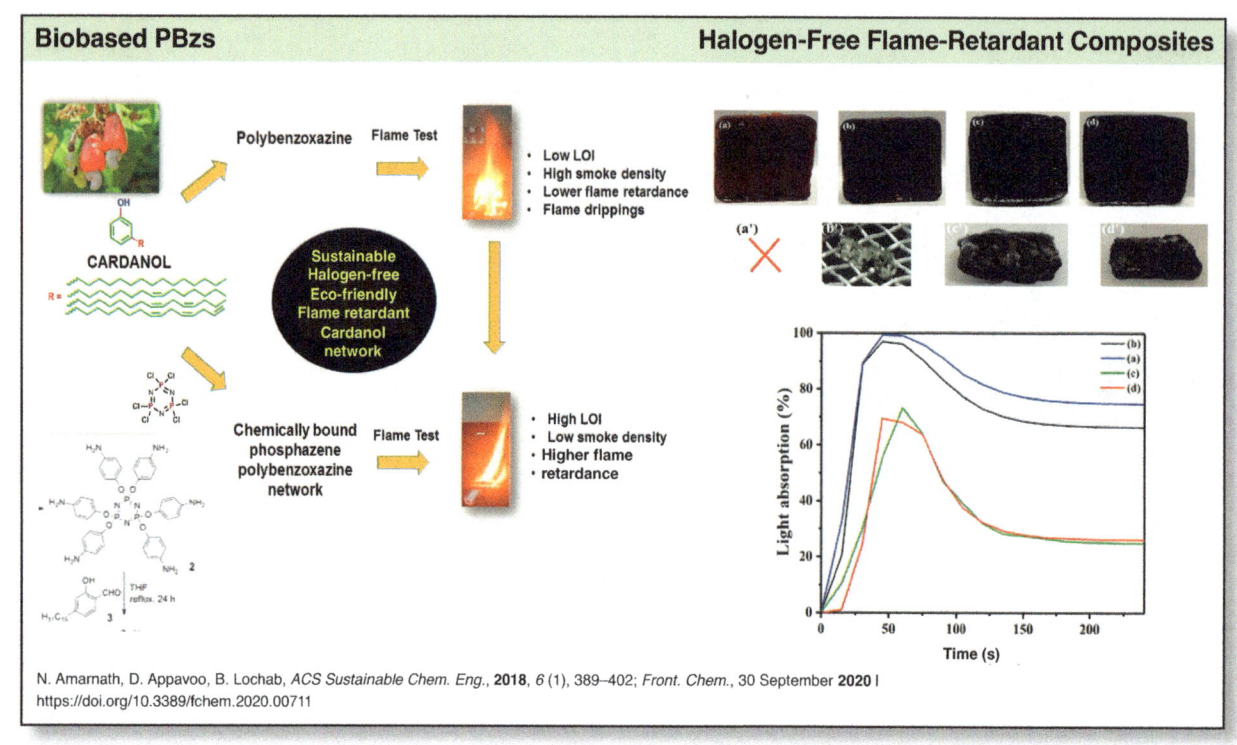

Figure 7

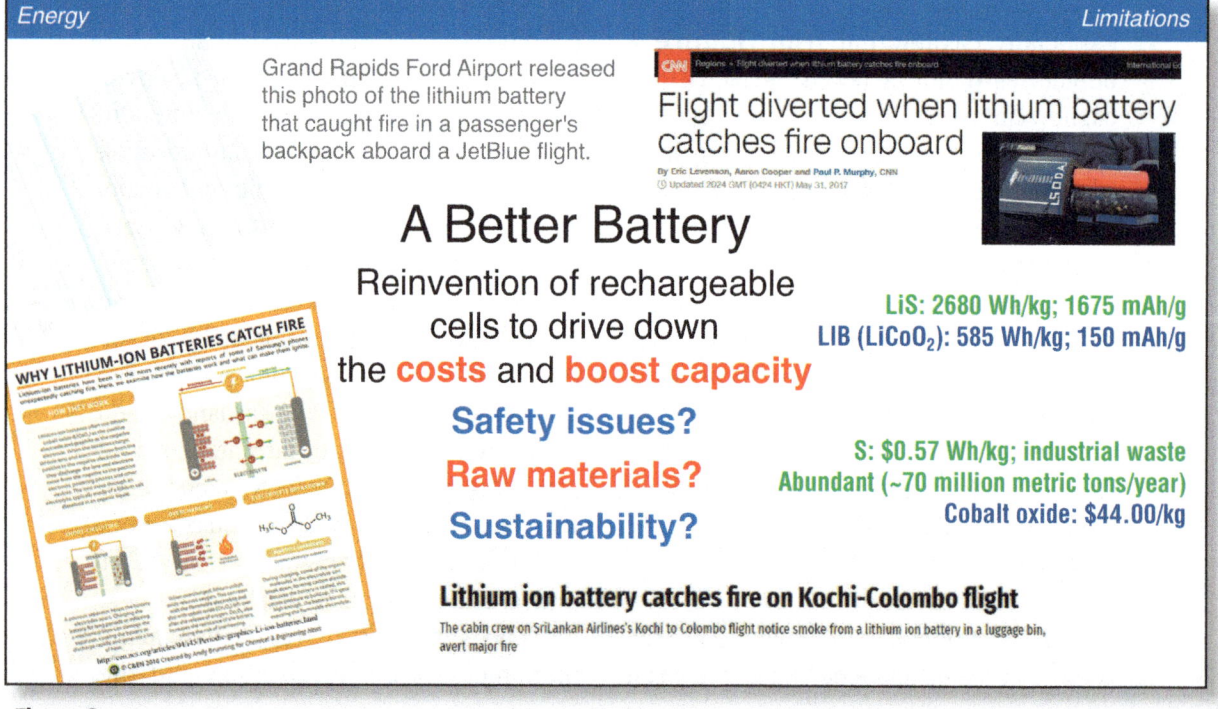

Figure 8

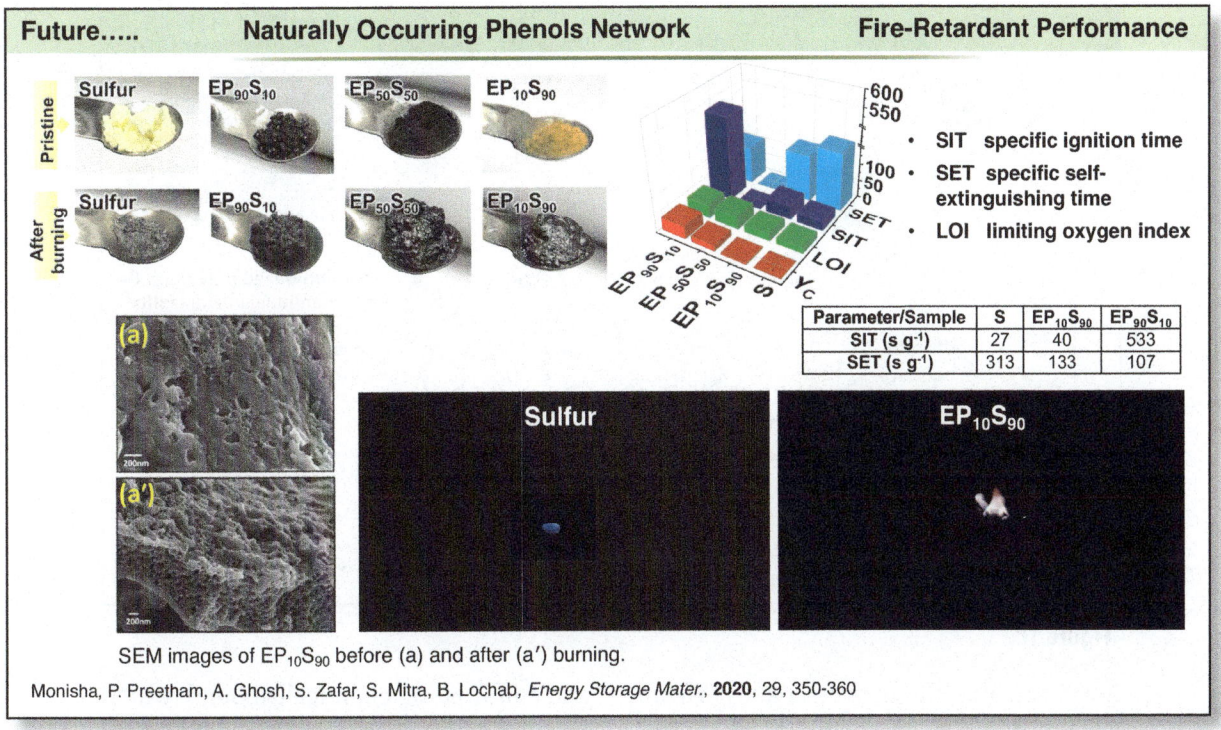

Future.....   **Naturally Occurring Phenols Network**   **Fire-Retardant Performance**

Parameter/Sample	S	EP$_{10}$S$_{90}$	EP$_{90}$S$_{10}$
SIT (s g^{-1})	27	40	533
SET (s g^{-1})	313	133	107

- SIT   specific ignition time
- SET   specific self-extinguishing time
- LOI   limiting oxygen index

SEM images of EP$_{10}$S$_{90}$ before (a) and after (a') burning.

Monisha, P. Preetham, A. Ghosh, S. Zafar, S. Mitra, B. Lochab, *Energy Storage Mater.*, **2020**, 29, 350-360

EP is Eugenol-phosphazene monomer.

Figure 9

respective applications being derived through the solventless loop constitute a much greener and more sustainable process. We are also exploring the polymer for manufacturing flame-resistant composites (**Figure 7**), and we are investigating their properties, which are well acknowledged in the scientific community. Thus, we are thinking locally but with a global perspective regarding this cardanol sourced polymers.

This local-and-global perspective also comes into play in our exploration of local materials to produce better batteries. Nowadays, lithium-ion batteries are in general use—yet they have a limited energy capacity (**Figure 8**). Since our energy needs are continuously increasing, we are called to discover superior alternatives and/or improved technologies.

In terms of finding alternatives to enhance energy sustainability, we are exploring elemental sulfur, which is obtained from the petroleum industry as an industrial waste. We are combining this with eugenol—extracted from clove oil, which is abundant in India—to make various copolymers. If we ignite sulfur (a component of lithium–sulfur batteries), it burns with a blue flame, as you can you see in **Figure 9**. We made a copolymer with this elemental sulfur, and as you can see, it produces a yellowish flame, has a very high ignition time, and is quickly self-extinguishing.

Now, if we made batteries from this polymer, they would have a robust tendency to mitigate fire, which is a strong advantage. We have done lots of explora-

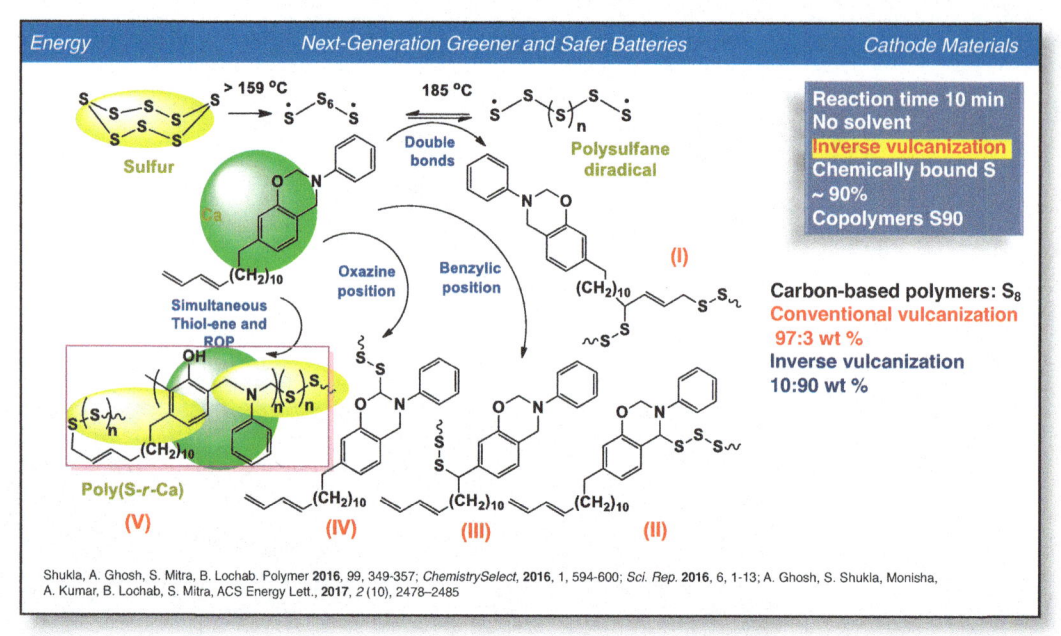

Figure 10

tion to identify the fire-retardant performance of this class of polymers, which has been very encouraging. This research has had in mind the next generation of batteries, which likely will be lithium with sulfur built into a eugenol–sulfur copolymer. This will be more typically important for tropical countries, where batteries catching fire will become a significant issue.

Again, we are using this kind of copolymerization technique, which has no solvent requirement, inverse vulcanization, and sulfur being chemically bound, which is a strong advantage in terms of battery performance (**Figure 10**).

With this, I would like to thank my research group, and as I always say, we have to look for better solutions for a sustainable world. If we can do the best job using waste, I think that is even more important. So phenolic compounds, as Hon. Mike Lancaster has stated, have a great future, and this challenge is being enthusiastically taken up by India. Thank you. ◊

References

Amarnath, N., D. Appavoo, and B. Lochab. 2018. "Eco-Friendly Halogen-Free Flame Retardant Cardanol Polyphosphazene Polybenzoxazine Networks." *ACS Sustainable Chemistry & Engineering* 6 (1): 389–402. https://doi.org/10.1021/acssuschemeng.7b02657.

Appavoo, D., N. Amarnath, and B. Lochab. 2020. "Cardanol and Eugenol Sourced Sustainable Non-halogen Flame Retardants for Enhanced Stability of Renewable Polybenzoxazines." *Frontiers in Chemistry* 30 September 2020. https://doi.org/10.3389/fchem.2020.00711.

Ghosh, A., S. Shukla, M. Monisha, A. Kumar, B. Lochab, and S. Mitra. 2017. "Sulfur Copolymer: A New Cathode Structure for Room-Temperature Sodium–Sulfur Batteries." *ACS Energy Letters* 2 (10): 2478–2485. https://doi.org/10.1021/acsenergylett.7b00714.

Ghosh, N.N., B. Kiskan, and Y. Yagci. 2007. "Polybenzoxazines—New high performance thermosetting resins: Synthesis and properties." *Progress in Polymer Science* 32 (11): 1344–1391. https://doi.org/10.1016/j.progpolymsci.2007.07.002

Lichtenthaler, F.W., and S. Peters. 2004. "Carbohydrates as Green Raw Materials for the Chemical Industry." *Comptes Rendus Chimie* 7 (2): 65–90. https://doi.org/10.1016/j.crci.2004.02.002.

Lochab, B., S. Shukla, and I.K. Varma. 2014. "Naturally Occurring Phenolic Sources: Monomers and Polymers." *RSC Advances* 4: 21,712–21,752. https://doi.org/10.1039/C4RA00181H.

——, M. Monisha, N. Amarnath, P. Sharma, S. Mukherjee, and H. Ishida. 2021. "Review on the Accelerated and Low-Temperature Polymerization of Benzoxazine Resins: Addition Polymerizable Sustainable Polymers." *Polymers* 13 (8): 1260. https://doi.org/10.3390/polym13081260

Monisha, M., N. Amarnath, S. Mukherjee, and B. Lochab. 2019. "Cardanol Benzoxazines: A Versatile Monomer with Advancing Applications." *Macromolecular Chemistry and Physics* 220 (3): 1800470. https://doi.org/10.1002/macp.201800470.

——, P. Permude, A. Ghosh, A. Kumar, S. Zafar, S. Mitra, and B. Lochab. 2020. "Halogen-Free Flame-Retardant Sulfur Copolymers with Stable Li–S Battery Performance." *Energy Storage Materials* 29: 350–360. https://doi.org/10.1016/j.ensm.2020.04.041.

Samak. G.R. Gangwar, A.S. Meena, R.G. Rao, P.K. Shukla, B. Manda, D. Narayanan, J.H. Jaggar, and R. Rao. 2016. "Calcium Channels and Oxidative Stress Mediate a Synergistic Disruption of Tight Junctions by Ethanol and Acetaldehyde in Caco-2 Cell Monolayers." *Scientific Reports* 6: 38899. https://doi.org/10.1038/srep38899.

Shukla, S., and B. Lochab. 2017. "Chapter 25: Lignin-Based Phenols: Potential Feedstock for Renewable Benzoxazines." In *Advanced and Emerging Polybenzoxazine Science and Technology*, edited by H. Ishida and P. Froimowicz. Elsevier Inc. 473–498.

——, N. Yadav, and B. Lochab. 2017. "Chapter 24: Cardanol-Based Benzoxazines and Their Applications." In *Advanced and Emerging Polybenzoxazine Science and Technology*, edited by H. Ishida and P. Froimowicz. Elsevier Inc. 451–472.

——, A. Ghosh, P.K. Roy, S. Mitra, and B. Lochab. 2016. "Cardanol Benzoxazines: A Sustainable Linker for Elemental Sulphur Based Copolymers via Inverse Vulcanisation." *Polymer* 99: 349–357. https://doi.org/10.1016/j.polymer.2016.07.037.

——, A. Ghosh, U.K. Sen, P.K. Roy, S. Mitra, and B. Lochab. 2016. "Cardanol Benzoxazine-Sulfur Copolymers for Li-S Batteries: Symbiosis of Sustainability and Performance." *ChemistrySelect* 1 (3): 594–600. https://doi.org/10.1002/slct.201600050.

Session 2 | Presentation Two

Rethinking the Plastics Revolution

Professor Michael Shaver
Polymer Science, The University of Manchester
Director, Sustainable Futures and Sustainable Materials Innovation Hub
Manchester, UK

What a challenging few years these have been. A global pandemic, unprovoked invasion and escalation to war, and climate change realities conflicting with unprecedented petroleum prices have all contributed to a cloudy future. However, the view that the end is nigh means it is ever more important and timely to come together and seek shared solutions to what are complex challenges. It is thus my pleasure to speak at this important Twenty-Eighth International Conference on the Unity of the Sciences (ICUS) and discuss some of our contributions to a more sustainable future through the lens of plastics.

While I was originally born in Canada, I now work at the University of Manchester in the UK. There, I wear several hats, including serving as director of the Sustainable Materials Innovation Hub and working in the Henry Royce Institute, the UK's national lab for advanced materials. However, while I was trained as a chemist, our team does not sit in a physical sciences silo. Instead, we bring together social scientists, economists, and environmental scientists alongside our core team of chemists and engineers to tackle the conflation of sustainability, plastics, and systems.

This is because plastics sustainability

About the Speaker

- Professor of Polymer Science; Director, Sustainable Materials Innovation Hub, Henry Royce Institute, The University of Manchester, UK.
- Researches the development of biodegradable materials and polymers with applications in medicine, commodity plastics, economy, and sustainability.
- Principal Investigator, One Bin to Rule Them All Project.
- Recipient of the MacroGroup Young Polymer Scientist Award (2015), Young Academy of Scotland (2014–2018), and Fellowship in the Royal Society of Chemistry (2018), among others.
- Editor, *European Polymer Journal*.
- PhD in Chemistry, University of British Columbia, Canada.

The Choices We Make

You can make, transport, use, and dispose of over 500 Styrofoam cups for the total cost of 1 ceramic mug

Figure 1

does not simply depend on the materials themselves, but also the choices we make as consumers. If we wanted to make the more sustainable choice for drinking a beverage, for instance, we could choose to drink it out of a ceramic mug or a polystyrene cup. However, which is more sustainable? When I deliver this presentation in person and poll the audience, the decision is near unanimous: Choose the ceramic mug. If we think about energy—and we have already heard several talks about net-zero goals—we can make, transport, use, and dispose of over 500 Styrofoam cups for the total energy cost of one ceramic mug (**Figure 1**). If we look at alternative materials, they are (usually) more energy intensive to secure the resource, manufacture the products, and transport them. Even reusing the mug costs significant energy: heating up the water, manufacturing the soap, and cleaning the water afterward. Plastics indeed save significant resources: They minimize packaging waste (by weight or number of items), decrease food spoilage, help create more lightweight vehicles that are more fuel efficient, and underpin modern insulation and lower energy costs.

The problem, however, is that this sustainability through their use does not translate into us treating them as valuable. Our identification of them as "waste" underpins our throwaway, single-use culture and leads to pollution at the macro, micro, and nano scales. Changing mindsets and creating value is hard, however, because of the complexity in what is quite a simple material. Take the oft-demonized grocery store. A package of meat may look like a "plastic package," but it is there for a reason. The first thing we

- **WHY?** Shelf life. Food safety standards. Not dyeing. **KEEP DOING!**
- **WHY BLACK?** Attractive meat tray drives purchases. **STOP DOING!**
- But it is not just one thing…
  - Plastic tray or plastic-reinforced paper tray
  - Plastic lamellar film (5+ plastic layers!)
  - Plastic adhesive / heat seal x 2
  - Plastic-reinforced paper label
  - Adsorbent plastic mesh soaked in meat juice (gross)
- **Design is essential, driven by practice not wishful thinking**

Figure 2

need to do is identify why the plastic is there in the first place. It increases shelf life, decreasing the carbon footprint of our food supply chain. It improves food safety standards, preventing sickness and death. Anything else that inhibits its value at end of life is what needs changing.

The biggest challenge is that the complexity of design makes recovery of value hard. The packaging of the meat in **Figure 2** appears simple but is in fact quite complex. It involves a plastic tray, a laminated film,

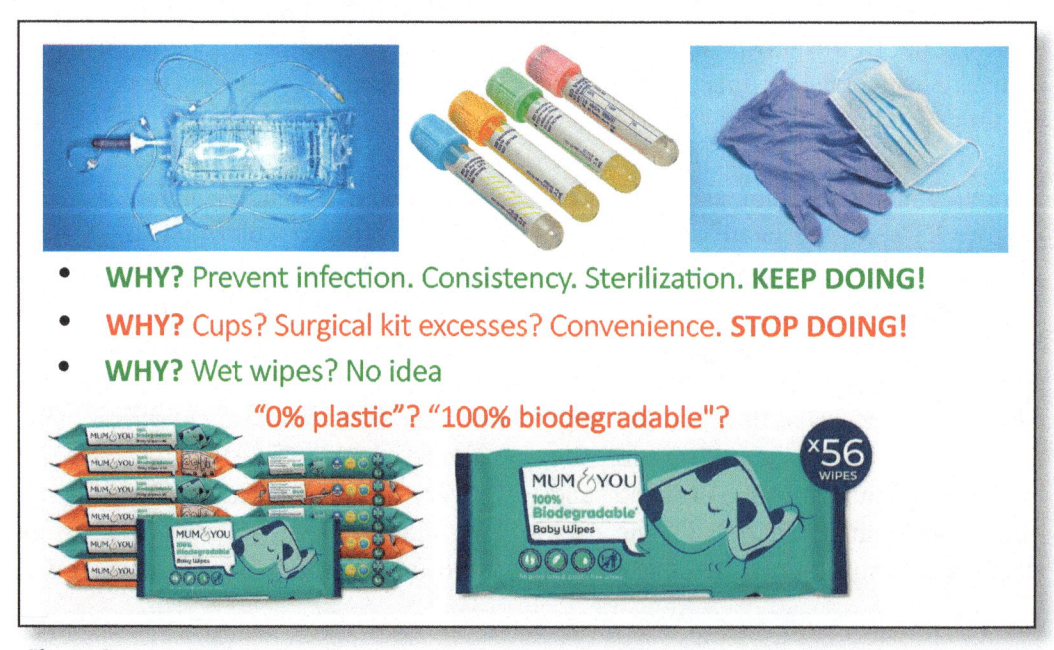

- **WHY?** Prevent infection. Consistency. Sterilization. **KEEP DOING!**
- **WHY?** Cups? Surgical kit excesses? Convenience. **STOP DOING!**
- **WHY?** Wet wipes? No idea

"0% plastic"? "100% biodegradable"?

Figure 3

adhesives, heat seals, and a plastic-reinforced paper label, and underneath it all is likely a plastic mesh soaked in meat juice. Each has a different composition and a different potential fate. That imagined fate is only realized by understanding social practice—we can likely get consumers to rinse out and recycle a tray, but no one is wringing out a mesh soaked in meat juice.

The same is true in any setting. The pandemic has highlighted our dependence on plastics to prevent infection, ensure sterility, and save lives. We need to continue these things, rather than risk uncertainty into surgery or spread disease. We must be mindful of how advertising intersects with social practice, where excessive greenwashing on product labels can conceivably influence consumers and harm the environment (**Figure 3**). For example, if the word *biodegradable* on a label makes people think they can litter without consequence, has it served its purpose?

Then, what does a sustainable polymer or plastic look like? I would argue that there is no such thing. Sustainability is necessarily relative—we must always look at what we are comparing things to. We must recognize that plastic products and the polymers they are composed of hold massive societal value. Alternatives (at least with our current energy supply) produce more waste and more CO_2. The diversity of plastics means that there are no panaceas and that significant optimization of each component is necessary.

It is not about one solution. We need to reduce the plastic that is unnecessary, reuse it where it is efficient and safe to do so, and recycle it—mechanically or chemically—in the most efficient way possible, and only then should we consider longer-term cycles or storage like composting or landfilling. All of this suggests that there is a naivete to any plastics-free world. We need a world that recovers value from all plastics. This is enabled by the *sustainable system* in which those plastics exist (**Figure 4**).

SO WHAT DO SUSTAINABLE MATERIALS LOOK LIKE?

Plastics (polymers) hold massive societal value; alternatives will produce more waste and more CO_2

Plastics are diverse beasts, requiring optimization!
REUSE? RECYCLE? DEGRADE? PYROLYZE?

Consequences for a plastics-free world?
What about a place that recovers value from all plastic?
SUSTAINABLE SYSTEMS, not SUSTAINABLE PLASTICS

Figure 4

If you remember one thing from this presentation, please let it be this: It does not matter if something is reusable or recyclable or compostable or biodegradable if it is not reused, recycled, composted, or biodegraded. We must start using past tense terminology to judge sustainability.

This is our challenge, so what approach must we take? We at the University of Manchester run a project called *One Bin to Rule Them All*. This project integrates materials science, shared business models, and social practice studies into an imagined future enabled by simplified, tagging-enabled plastic waste management. The project works with partners from across the supply chain to explore the policy and infrastructure levers that enable plastic to be held in its highest-value condition, integrating and choosing among multiple potential fates. We currently sort plastic based on its polymer backbone. For example, a polyethylene terephthalate (PET) bottle would get sorted alongside a PET tray. Some of those trays are lined with a different polymer, like high-density polyethylene (HDPE). This leads to a lower-value recyclate because of the diminutive nature of the sorting process. What has been interesting to discover is that the solutions needed are not technocentric—we have systems to deliver this change now, but it requires agreement across supply chain actors who are not used to working together.

This project is tagging agnostic. There are lots of different ways to tag materials—embedded marker molecules, printed QR codes, embossed codes, flexible semiconductors. They hold different amounts of information, and indeed can be used to make different decisions. The key for our work is to understand the decisions that first could, and then should, be made. The decisions are mapped onto economic sustainability pathways, based on both our current infrastructure and potential investment (**Figure 5**). The deci-

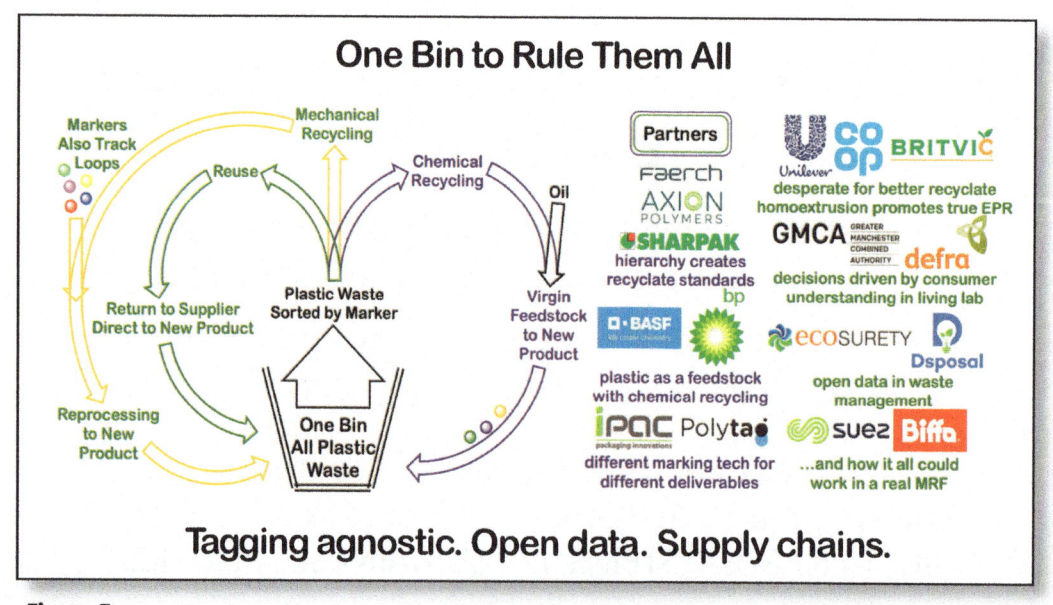

Figure 5

sions change as the infrastructure investment changes, meaning that we also see great geographical differences.

As an example of this materials hierarchy, how would we best sort to prioritize value? Some decisions are made to avoid harm: Products could be labeled to ensure they do not pollute the recycling stream. With current infrastructure, this could be medical plastics, biodegradables, aseptic cartons, and the like. The resulting product stream could then be segregated into such categories as multi- versus monomaterials and food versus nonfood to create value. For a specific plastic stream, value is different if segregation is made for bottles versus trays; the materials are often fundamentally different.

Is further segregation worthwhile? Colored versus natural? Sorting based on additives or melt flow indices? What fractions should go to mechanical versus chemical recycling? Will this decision change with upcoming legislative changes? In the UK, the Plastic Packaging Tax and the new Extended Producer Responsibility environmental regulations are having a major influence on industry practice—but legislation and public pressure vary widely from country to country and region to region.

Similar challenges exist when looking at new polymers. Our lab has a long history of developing new motifs in biodegradable polyesters. We pioneered the use of dioxolanone monomers to synthesize, in high yields, polymers that (a) are entirely derived from renewable feedstocks, (b) can tune functionality to deliver a range of thermal and mechanical properties, and (c) are circular, in that

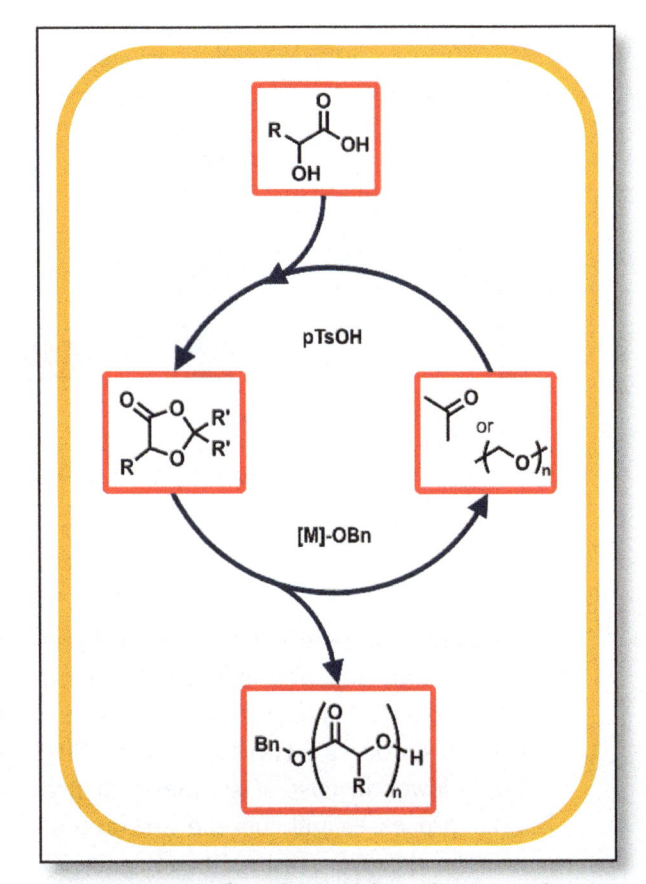

Circular economy for poly(mandelic acid) and other functional polyesters. **Figure 6**
Cairns, S.A., A. Schultheiss, and M.P. Shaver. 2017. "A broad scope of aliphatic polyesters prepared by elimination of small molecules from sustainable 1,3-dioxolan-4-ones" *Polymer Science* 8: 2990–2996.

they are readily depolymerized back to their parent monomers (**Figure 6**).

However, a challenge becomes apparent when you look at the reality behind the curtain of this positive image. Take poly(mandelic acid) (PMA), for instance, where the stereoregular form (isotactic) serves as a biodegradable mimic of that same polystyrene used in the Styrofoam cup mentioned at the beginning of this paper. We can make this efficiently and

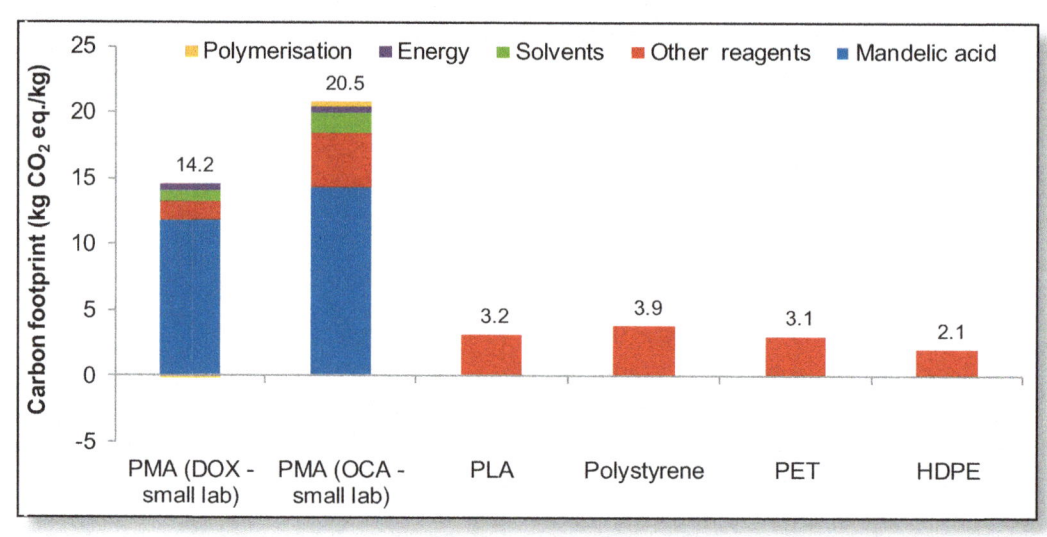

Figure 7 — **Comparing carbon footprints** of poly(mandelic acid) from dioxolanone and O-carboxyanhydride monomers with common packaging plastics.

economically in our lab, in a fully recyclable system, and obtain polymers with Tgs > 110 °C (essential for coffee drinkers!). Our lab's process is two hundred and fifty times lower in cost of synthesis in previous processes and cuts the carbon footprint of the process by 30% (**Figure 7**). But what are we comparing this system to? If we look at those numbers versus polystyrene, we realize that this new "best" route to PMA is *3.6 times higher* than for polystyrene! Context really does matter, as does transparency across the industry.

We have the privilege of being an academic group that celebrates truth over sales. The biggest challenge is the contextualization of these plastic challenges within the broader environmental sustainability challenges we face. For this reason, The University of Manchester (UoM) has established the Sustainable Futures research platform. Sustainable Futures brings together the unique depth and breadth of internationally leading research at UoM to produce integrated and truly sustainable solutions to urgent environmental challenges. Our research experts are helping to build a greener, fairer, and healthier future for all. Interdisciplinary collaborations, cross-sector partnerships, and pioneering discoveries benefit the environment, economy, health, well-being, and culture, all globally.

This enterprise is necessarily interdisciplinary. We aim to align, connect, celebrate, and then grow our exceptional and diverse research on environmental sustainability. Sustainable Futures has two equally vital objectives:

- to enable and catalyze the internal and external connectivity to deliver change that is meaningful to partners in government, industry, and communities

- to showcase our world-leading strengths in environmental sustainability to promote concerted and sustained change.

This new platform represents both a celebration and an ambition. It is a recognition that we are already world leading in tackling—and, importantly, in helping other organizations tackle—global challenges in environmental sustainability. It is also a recognition that the urgency and complexity of these challenges continue, and that we have an opportunity to grow our global leadership in transforming our world for the better. The challenges of environmental sustainability are complex and necessarily interdisciplinary. The interrelationships provide for a safety net to avoid solutions in one space creating unintended consequences in another, especially when economic and social values are also considered. It is important to be explicit in this regard: While our challenge is to address environmental challenges, we address those challenges from the combined perspectives of social, economic, and environmental sustainability.

Sustainable Futures' upcoming year has an internal focus on supporting our strengths, growing our understanding, and building our connectivity. We aim to develop the UoM sustainability community in a way that aligns with the university's strategic focus through the lens of the following five challenges:

- *Resilient futures:* Considering increasingly disruptive environmental change, how do we build a more resilient society while ensuring that our natural environments are protected, restored, and sustained?

- *Resourceful futures:* How do we sustainably manage our finite resources and recover value from, rather than lose materials into the environment?

- *Net-zero futures:* How can we deliver the rapid emissions reductions required for a sustainable future in a socially and economically sustainable way?

- *Healthy futures:* How do we manage risks to humans and ecosystems while making and prioritizing environmentally sustainable transitions?

- *Inclusive and prosperous* futures: How do we deliver environmental sustainability that is fair and just and enables thriving, diverse communities?

Our success in delivering social and environmental impact (for 2022, we are ranked number nine out of more than 1400 universities in the Times Higher Education Impact Rankings in terms of action on the UN's Sustainable Development Goals) is bringing more focus on our efforts.

The plastics challenge, if viewed in isolation, will lead to unintended consequences. It is only when they are viewed in a systemic way that we can come to deliver holistic solutions. The corroborations you hear between the diverse talks at this ICUS event are important, but the contestations—the pushes and pulls that work against each other—are the true opportunities for holistic solutions to not only eliminate the release of plastic waste but extend the time that humans have the privilege to live on this wondrous planet. ◊

Session 2 | First Commentary
on the Presentation by Professor Michael Shaver

Rethinking the Plastics Revolution

Professor Mark Miodownik
Materials and Society, Department of Mechanical Engineering,
University College London, UK

I would like to thank Prof. Shaver so much for his great talk. I really enjoyed it. I want to emphasize that the systems approach to environmental issues like plastics is just so important. His example of whether to choose a plastic cup or a ceramic mug was a powerful example that we should all think about deeply, because having reusable drinking vessels seems like an obvious part of the solution to the plastic waste problem—deceptively so.

Looking at ceramic cups, for example, we really have to accept the fact that they, just as with the plastic type, do not have a zero environmental impact, considering the energy use involved and the sourcing of all the other products that make a ceramic cup. All of these elements have an environmental impact, and when you look at reuse models for cups in different scenarios—and there are many companies now supplying ceramic cups or other types of vessels for reuse instead of disposable plastic cups—what you find when you do a life cycle analysis is that the loss rates really determine whether reusability is a better environmental solution or not. If a company gives people reusable cups in a large office environment, and you

About the Speaker

- Professor of Materials and Society, Department of Mechanical Engineering; Director, Institute of Making, University College London, UK.
- Researches animate matter, plastic waste innovation, and sustainable manufacturing.
- Fellow, Royal Academy of Engineering.
- Author of *New York Times* bestseller, *Stuff Matters: Exploring the Marvelous Materials That Shape Our Man-Made World*.
- Chosen by *The Times* as one of the 100 most influential scientists in the UK (2010).
- PhD in Turbine Jet Engine Alloys, Oxford University, UK.

end up losing 10% or 15% of them, or sometimes 20% of them, that loss actually can exceed the environmental impact of using disposable cups and recycling them.

Therefore, it is good to be wary, as Prof. Shaver says, of the unintended consequences of having just a blanket attitude, telling ourselves, "Well, we've got this plastics problem and so we will just replace plastics wherever we can." The issue is actually much more difficult than that. The plastic waste problem is a so-called wicked problem. We must not stumble into the attitude that we are just going to replace plastics with other materials, because no material has zero environmental impact. In the end, there is no such thing as a sustainable material. There is only a sustainable *system*. That system involves how you make a material and all the substances and products going into it, how you use those materials and those products, how you then reclaim them and reprocess them back into products again. That all requires a case-by-case analysis.

Prof. Shaver gave the example of the complexity of the issue in regard to food, where it is not that obvious, actually, whether you should get rid of some plastic packaging. As he said, that plastic reduces food waste and therefore all the CO_2 emissions and water usage that have go together with that.

Likewise, with single-use plastic medical products, it is of course vital to utilize them to make sure an operation proceeds successfully and that there is no pathogen contamination. However, let us take the example of medical gloves, the use of which is just huge in many hospitals across the world. They are simply overused in many cases compared to what is actually needed to achieve good health outcomes. Indeed, the use of medical gloves could be reduced by probably 50%, cutting costs and waste by assigning them for certain procedures where they are absolutely required and not having them in blanket use in every medical circumstance.

Another good example, one that we have worked on, is masks, and the COVID epidemic has shown us that we do not need to be using disposable masks everywhere. Certainly, we have shown that *reusable* masks, even in hospital environments, get the same health outcomes with much lower environmental costs.

Therefore, redesigning systems, as Prof. Shaver says, is thinking about the complexity of what we are replacing materials with, looking at all the possible unintended consequences and benefits. We must always think of end of life whenever we consider replacing something or bringing some new product into the world. What is the product's end of life? Do we have a plan for it? What is going to happen when it is finished? If we do not have a plan, we really need one. That is where I think Prof. Shaver's project called the *One Bin* is so ingenious, because really, if we think about it, we have got to have an approach that will be easy for a person, in a world full of "stuff," to be able to navigate where things should go at their end of life.

In each home, there are waste bins, and we need to know which bin to put things in. We do not want to make that too difficult for people. So dry things, things that are not wet or food contaminated, ideally we just let them all go in one container—the One Bin Prof. Shaver talks about—and then we can let robots later on down the line do the sorting, which they are good at. I think that is a huge opportunity, and it is surely the way we are going to go.

But then there is the wet bin in the home, which is the destination for food waste, feminine hygiene products, diapers, and the like. There should be a very simple way of dealing with those too, right? You want the citizen, the homeowner, the person who is involved here to be just completely clear about where waste should go. Let us say there is going to be a wet bin and one dry bin—that is, Prof. Shaver's One Bin and then the wet bin. Now, what should the materials destined for each bin be made of? Are they going to be disposable? Should they be biodegradable polymers? How do we navigate that?

There is the sense in society that if something biodegrades it must be good, but as Prof. Shaver has shown, that is not necessarily true—it is actually again really complicated. For example, if we have a system for collecting biodegradable polymers, and we know we can process them at end of life by collecting them from the home or hospital, taking them to a composting or processing site, and then turning them into a nutrient product that can fertilize the land to grow more crops, then, when you do the life cycle assessment, you may well find that biodegradable polymers are worth it.

However, this is not always going to be the case, and certainly in the case of takeout food packaging it is not the case. We should be quite careful about the idea that anything that is disposable and contaminated should be made of biodegradable polymers. That does not seem to be true. Overall, I think Prof. Shaver painted the right picture, which is that we have a lot of the tools necessary to make a huge impact. *Reduce* is always going to be the easiest win. *Replace* is a much more complicated choice. With every material and product, we need to really think it through case by case, looking at the energy involved and all the other environmental impacts. Thanks for inviting me to offer this commentary. ◊

Session 2 | Second Commentary
on the Presentation by Professor Michael Shaver

Rethinking the Plastics Revolution

Dr. Carly Fletcher
Research Associate, Department of Natural Sciences
Manchester Metropolitan University, UK

Distinguished chair, speaker, and guests. I would like to thank the ICUS XXVIII committee for this opportunity to provide a commentary on Prof. Shaver's paper "Rethinking the Plastics Revolution." I fully agree that times have been challenging, and now we must come together to determine what a sustainable future looks like and what role (if any) plastics have to play. To that end, this commentary will first explore plastics in modern society and the associated problems. Then it will ask, how viable are "sustainable plastics" as a solution? Can One Bin really rule them all?

Plastics are omnipresent in modern society. Prof. Shaver highlighted their role in keeping food fresh and in strengthening healthcare. If we take a snapshot of everyday life, plastics are everywhere. In the places where we work or study, plastics are found in stationery and electronic devices, in that well-earned takeout coffee, and in the home, where they occur in utensils, textiles, appliances, and toys. Even during social events, plastics contribute to safety, convenience, and style. Plastics are durable, adaptable, and inexpensive, and it is these attributes that have made them dominant (**Figure 1**).

About the Speaker

- Research Associate, Department of Natural Sciences, Manchester Metropolitan University, UK.
- Researches the safety and sustainability of biobased and biodegradable plastics, waste management policy, circular business models, and the transition to a circular economy.
- PhD in Environmental Management and Sustainable Development, Manchester Metropolitan University, UK.

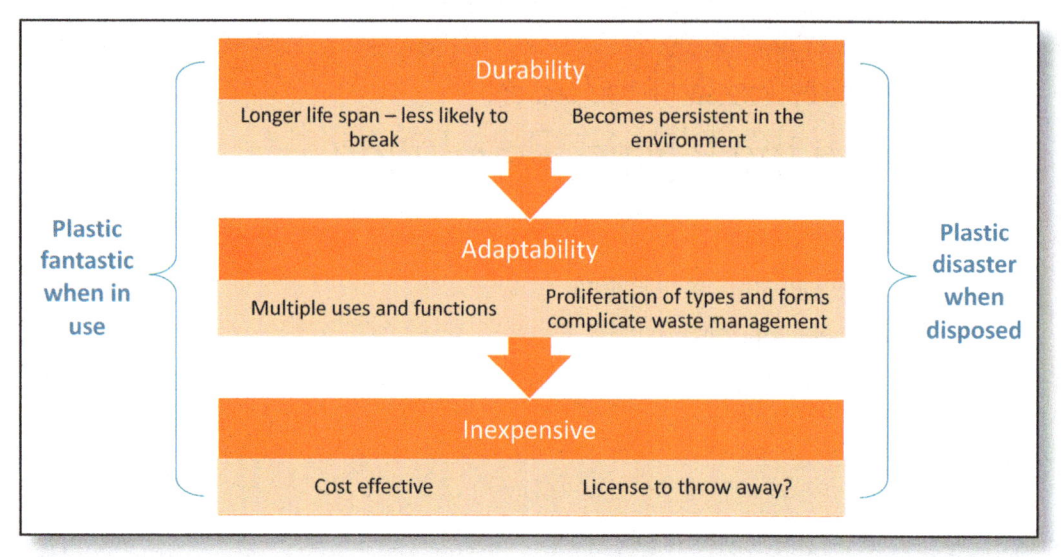

Figure 1   **Attributes of plastics:** two sides of the same coin.

Durability means that the products span a longer life and suffer less breakage. Adaptability has led to numerous uses, applications, and functions. Being relatively inexpensive, plastics are often cost effective.

However, these same attributes are also problematic. Because of their durability, plastics can become persistent in the environment, be that in the oceans, rivers, or countryside. Even in landfills, they remain long after disposal. Adaptability has led to proliferation of types and forms, which has implications for effective waste management. Finally, being inexpensive means that society as a whole does not value the intrinsic cost of plastic, thus driving a single-use, throwaway culture. As Prof. Shaver suggested, it is not the use of plastic that is the problem; the problems come from what we do with it afterward.

Addressing this complex issue, Prof. Shaver mentioned sustainable plastics and their limitations. This topic I will address from my current role within the Bio-Plastics Europe project (funded by the EU Horizon 2020 program). While focused on biobased and biodegradable plastics, the points are applicable to other materials. Can sustainable plastics solve this problem? No, they are not *the* solution, not a silver bullet, if you like. However, they could be part of a wider, more holistic effort to address the problem. Industry acknowledges this contribution and predicts growth in sustainable plastics, driven by three factors (**Figure 2**).

First, uptake is driven by policy. In the EU, the Green Deal and 2018 Plastics Strategy have acknowledged the role of alternatives, seeking to understand all the potential ramifications. Internationally, the UN's Sustainable Development Goals and the Ellen MacArthur Foundation's Plastics Pact Network also recognize their potential. The second driver is research, development, and innovation. Indeed, the development of new materials presenting

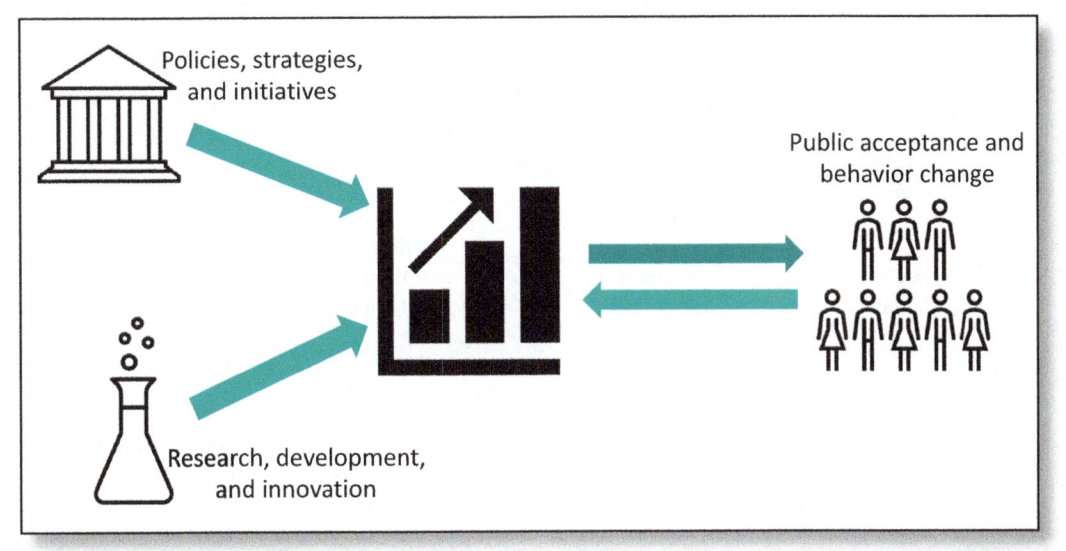

Viability of "Sustainable Plastics"   **Figure 2**

superior qualities or improved resource efficiency can also drive the market. The final piece of the puzzle combines both push and pull factors. The end user, who may passively consume or intentionally seek out sustainable products, can publicly accept them as the new norm. They may also raise the bar regarding expectations of performance. Conversely, a lack of awareness, unrealistic expectations, and entrenched behaviors can present barriers.

Taking all of this into account, initial recommendations from the Bio-Plastics Europe project include the need to combat consumer confusion and unrealistic expectations. Take, for example, the 100% biodegradable wet wipes highlighted by Prof. Shaver. Given enough time, most materials are biodegradable—even conventional plastics will eventually break down, given hundreds of years. Here, more specific terminology, such as *compostable* (preferably, distinguishing between home and industrial) is needed to manage expectations. On a related note, care must be taken not to create negative feedback loops. Just because a product can biodegrade does not give the user license to litter, and the use of sustainable plastics should not challenge efforts to reduce overall consumption.

Considering utilization, any substitutions should be appropriate. For example, the use of compostable food packaging seems sensible, as the packaging can be composted (residues and all) with other organic wastes. In addition, any producer claims should be transparent and clear, based on robust evidence. Finally, considering the entire value chain is essential. Focusing on biobased plastics, what and where the feedstock comes from has implications for sustainability. Producers also need to consider what happens to their product once disposed—which nicely links to the One Bin idea presented by Prof. Shaver.

To summarize, the One Bin to Rule Them All concept seeks to tag materials

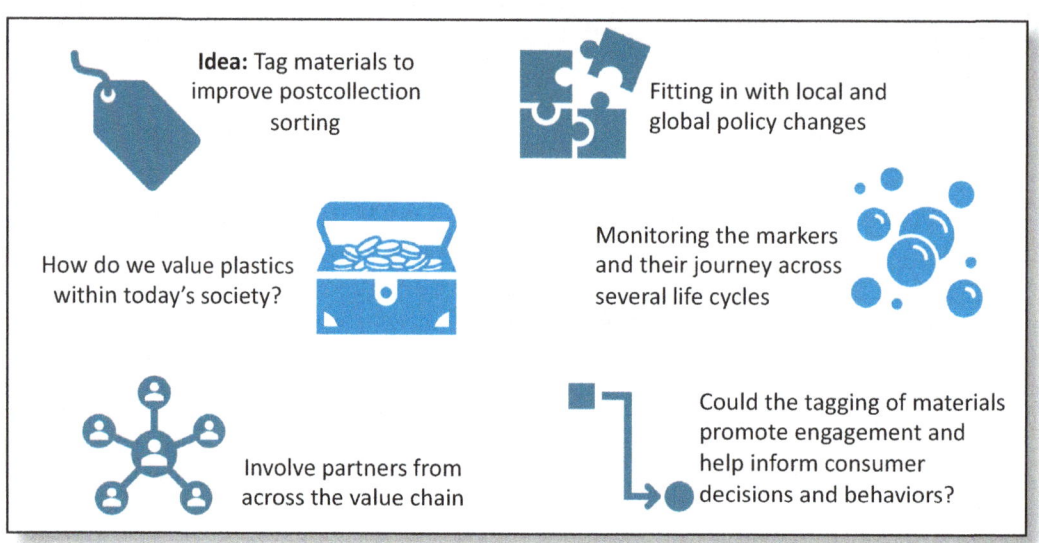

Figure 3 Can One Bin rule them all?

- **Idea:** Tag materials to improve postcollection sorting
- Fitting in with local and global policy changes
- How do we value plastics within today's society?
- Monitoring the markers and their journey across several life cycles
- Involve partners from across the value chain
- Could the tagging of materials promote engagement and help inform consumer decisions and behaviors?

in order to improve post-collection sorting (**Figure 3**). It also ruminates on the dilemma of how plastics are valued by society. With support from across the value chain, the project has the advantage of hosting a wide range of partners, including some big movers and shakers. While the implications of changing legislation have been noted, I wonder how this system could work with deposit/return schemes and standardized bin collection, both of which are forecast in the near future. Furthermore, I would be interested to hear how the markers could be monitored across different life cycles. A final thought would be to explore the possibility of expanding the use of the tags (particularly QR codes) to promote public engagement and help inform consumer decisions.

To conclude this commentary, I would like to extend my compliments to Prof. Shaver on the creation of the Sustainable Futures research platform. Looking forward to how we as a society envision and implement a sustainable future, I think you will agree that the contribution of all parties is required. To that end, I am pleased to see that Sustainable Futures has rejected the old ways of working within individual silos and instead endorses the more collaborative, interdisciplinary, and holistic approach, considering the impacts of the problem and potential solutions (including risks, opportunities, and implications) from across the full value chain. ◊

Session 2
Discussion

Michael Stenstrom (Chair): We are going to have about 30 minutes of questions and answers now. But this is the age of Zoom. So, we have to do it a little differently. You can see there is a "raise your hand" at the bottom of the screen. Therefore, if you have a question, we would like for you to click on that. It will appear on my screen, and I will then call on you. After I call on you, I would like for you to state your name and a sentence of where you are from—not as a resume but just a little bit—and then please say to whom you would like to address your question among our four commentators and two speakers. Okay, let us begin. Who would like to ask a question? Let me start. Oh, no, we have a question. Prof. Clark, did you want to speak?

James Clark: No, no.

Stenstrom: Okay, well, I would like to ask a question. It relates to the issue of source separation and dividing materials into bins to facilitate their recycling. Here in Los Angeles, we have three bins that we pick up in the city. They are a black bin, which is commonly called garbage; a green bin, which is for plant material; and a blue bin, which is for plastics, but which winds up containing lots of other things. We have had a pretty good response from the public in support of such separation. But it took leadership by the city and a lot of effort. And my question is, to the folks in the UK, how well does that work there? I suspect it actually works better. But I would like for you to comment on how well source separation works in the United Kingdom. Maybe Prof. Clark, since I see you right in front of me, you could start.

Clark: From my experience in terms of general waste. Do you mean in terms of household waste?

Stenstrom: Yes, household waste.

Clark: I think people on the whole, it does vary depending on which area of the country you live in. I am currently in Wales, and in Wales, it is very well separated. I have got a whole series of different bins in the kitchen area here for disposal. Back in Yorkshire, where I come from, it is a little less separated. I think we have only two alternatives there for separating our waste, and it does vary depending on where you are. I think people on the whole are pretty responsive. It has become part of their tradition, one of their habits now, and they are separating in the home. I would say the response is pretty good. On the whole, people do not complain about it as much as they used to.

Michael Shaver: I can perhaps give us a bit of commentary on that. I think what is interesting when you actually look at sort of the more complex social science studies on this, practice varies a lot, not based on just the willingness to do it, but based on where you are. The number of bathroom bottles, for instance, that you put into the recycling bin, those being your shampoos and that kind of stuff, is dramatically less than food-based bottles. The convenience of practice and the automation is really what then leads to a higher amount of material coming through.

I think it is really important to look at two things: the volume of material, which is created by consistency of collection, and then the quality of that material in terms of how you actually segregate it once it comes through. What the work in the UK has done has to focus on improving the quality of material by collecting less. In many places, authorities will just collect bottles, or you will have a deposit/return scheme that controls feedstock. There are real societal implications and financial implications to doing that. But that is very different from trying to optimize for a large volume and a large quantity, which is where the system breaks down.

Stenstrom: Thank you very much for that. We have a question from Arta Musaraj, so please give her the floor.

Arta Musaraj: Hello, and thank you to our speakers and discussants. The utility of this session is obviously radical to us all. I am Prof. Arta Musaraj, the editor in chief of *Academicus International Scientific Journal*. We shared with Prof. Stenstrom in 2020 at the same conference, and it is now a chance to greet you. Well, I had two questions and two comments at the same time. The first one is to what extent the raw materials and production processes are incorporated into the economic concept of utility and use of a product or service by being a social concern in this way. This is considering that the chemical industry is found often in the first chain or in the sector that produces the raw materials.

The second question and comment deal with the capital investments, considering here the production systems. Has there ever been a calculation of the cost of shutdowns and restarts? By restart here, I mean a restart with green chemistry and including both social and economic costs. This is because I noticed that the term *desalination technology* was used. We should consider the structure of the industry—not just that of the chemical one but all the production chains, production services, situations, or systems that produce the goods and services we use in our daily lives. I would like to address the first question to Hon. Mike Lancaster and the second to Prof. Michael Shaver. Thank you.

Discussion

Mike Lancaster: I am sorry, but could you just repeat the first question? I could not really understand what it was.

Musaraj: Yes, thank you. It will be my pleasure. To what extent are the raw materials and production processes incorporated into the economic utility of products? Here I mean the economic concept of utility and usage of a product or service by being or causing a social concern. This is considering that the chemical industry is found in the first chain, or in the sector that produces the raw materials. Of course, it is the case that the interdisciplinary methodology is the only one to be used in deciding, helping us in choosing, or creating an equilibrium between choices, effective ones and effective to the lives of all of us and our needs. Thank you.

Lancaster: I mean, around 95–98% of everything we consume contains manmade chemicals of some description. I am not really sure what you mean by *utility*. In terms of chemical processes, they differ very widely in terms of how much of the raw materials get incorporated into the finished product. It depends on the complexity of the processes. But I did not really understand your concept of utility.

Musaraj: Utility in economic terms. I mean utility to the user or person who consumes a good or service.

Lancaster: Most of the chemical industry is several steps away from the end consumer, so most of the chemical sector sells to itself and is normally three or four steps from something appearing on a supermarket shelf. It is the consumer-goods fabricators that deal more with the end consumer, so I am not really able to answer your question. I do not know if anyone else from the panel could better answer it.

Clark: A comment that may be worth making, Mike, is the fact that through green chemistry the drivers of change are often coming at the end of the supply chain. It is often the consumers, retailers, or consumer-goods producers that are demanding green chemistry. It is kind of interesting, because they are the ones that are basically demanding more and more sustainability in the products they use. They are more concerned about issues like carbon footprint and environmental footprint.

In a way, we certainly found in research, that they are the ones who are kind of encouraging the green chemistry, and they are persuading their suppliers, and those suppliers are persuading their suppliers. And so it works its way back up the value chain into the chemical manufacturing industry, and of course, ultimately, to raw materials production as well. So, there is certainly a big pull, I would say, coming from the consumer end of supply chains involving all sorts of chemical products.

Stenstrom: Well, thank you. That is a really interesting question. We have another question I would like to get to from Jin Choon Kim. Please go ahead.

Jin Choon Kim: I would like to raise one question for Hon. Lancaster. I really enjoyed your beautiful presentation. This question is kind of a fundamental question. You really emphasize green chemistry. Chemistry usually has molecule levels, and I think your green chemistry is based on polymers, or chemical ions and chemical compounds. In that case, is there any possible way to consider quantum entanglement by quantum physics, that being the quantum entanglement of molecules or holistic features of molecules, or even implicating order or hidden order among molecules? If we consider this more internal aspect over the physical aspect, then maybe we will find some different way to reduce pollution, degradation, and toxic by-products. I would like to hear your opinion regarding this. Thank you.

Lancaster: Yes, that is a very interesting comment. I am sure you are absolutely right, and as we start to understand a lot more about what is inside molecules and how we can influence the very diverse range of reactions possible, I am sure we will come up with much better ways to do synthetic chemistry. And, yes, very interesting pathways that we have not even yet thought of could possibly be developed, which will reduce waste and produce new kinds of materials. I think a great understanding of the physics as well can really help in the future. I think there have been some recent developments in that process, where people are able to predict what happens at the molecular level through a greater understanding of the smaller particles inside the atom. I think it is a very fascinating topic.

Stenstrom: Well, we do not have a question on the floor at the moment, but a topic that was in the material and leading up to the conference, which we could discuss and I think is worthwhile, is biodegradable plastics. I can remember when I was a graduate student, we talked about biodegradable plastics. And some of those plastics were not really biodegradable. They just disintegrated into small particles, which were worse perhaps in the environment than large plastic objects would be. I can remember, when I toured Antarctica some years ago, I saw a picture of a krill, somewhat like a shrimp but not related, and it has a transparent body. They held it up to the lens and we could see microplastic particles inside the krill's body. We have experts here on plastics and perhaps biodegradable plastics, so why do we not have someone comment? Maybe Mike Shaver, would you like to start?

Shaver: Yes, sure. And Mark can then tell me how I am wrong. I think that what is important is to recognize that biodegradation—and Carly talked about this—is a process that happens over lots of different time scales. The evaluation of whether something degrades and what it degrades into has to be dealt with in the environment of release. When we have uncontrolled release, that makes that really difficult. Regardless of what our imagined fate for those materials is, we have to be able to have a system that enables the segregation of the material to get it into that past-tense terminology. If we want to go down the biodegradation route, that still requires infrastructure to be in place to be able to ensure that that fate exists.

That does not mean that if something ends up in the ocean—where you have colder temperatures, a high saline environment, and lots of environmental reasons that degradation would slow down—it would not lead to microplastics. Therefore, the source of those microplastics actually might have a real impact there. We have done a little bit of ecotoxicological studies on zebrafish to say that if I feed them a bunch of microplastics, which are made of HDPE, nothing happens. And if you do the same thing with biobased materials, that being food, that can lead to changes in those organisms' development. Thus, the number of unknowns in that space really means that we need to understand both the fate of those materials and what will happen to them from an ecotoxicological standpoint in the array of different environments in which they can be released.

Stenstrom: Yes, that is certainly true for the array of different environments. Does anyone else want to comment?

Mark Miodownik: I will come in. I do want to let Carly speak, if you want to just jump in there as well.

Stenstrom: Go ahead, Mark.

Miodownik: Okay. We have been doing field studies on biodegradable plastics to see what happens when they escape the system that they are designed for. Of course, you can see quite quickly that if the conditions are not exactly right, they behave very much like normal, conventional plastics—that is, they basically persist in the environment. The microplastics issue is really interesting, because essentially, even when they do go through an industrial composting system and come out the other end as compost, the residency time is not clear yet for the microplastics that are residues in that compost and that end up back on the land.

In the UK, anyway, and, as far as I know, in Italy, Germany, and France, there is a huge degree of worry now about the fact that we do not really understand yet the fate of the microplastics from biodegradable plastics, even when they go through composting systems. And that is to set aside the other problem that Mike talked about, which is that when they do not go through those, they are completely uncontrollable. I think it is a big scientific question mark about what we want from biodegradable plastics and which products they are the best solution for. I would say that—even though we are all doing work in this area—I do not think any of us could really hand-on-heart say that there is this class or product that these biodegradable plastics are the best for.

I do not think any of us could say that at this point. This is not to say that we should not all be still working on that, as we are. But I would just, as Mike has just done, sort of argue for caution. I know that in the American market and in other markets worldwide there is quite a lot of enthusiasm for this product type, driven, I think, by shopper and customer demand for a more sustainable future. I think we have got to discourage companies from jumping on this bandwagon until we have a system that works.

Stenstrom: Thank you very much for that. My own take on biodegradable plastics is that biodegradable by itself is not sufficient. We need to know how long it takes and under what conditions plastics degrade, because biodegradable plastics going to a treatment plant are processing in hours or just a few days. Biodegradable plastics going into composting might be months. Thus, we really need more technology there.

So, it is coming up on 2:58. We have two more minutes. Who would like to ask another question or make a comment? I do not see any. And by the way, Mark, you have a great clock behind you and we can see that it is 10:58 where you are. That is great. So, I think we have a hand up. Bimlesh Lochab, please go ahead.

Bimlesh Lochab: Thank you very much. I am just wondering about the nicely stated, "realign, connect, celebrate, and grow together." Now, the scenario in developed countries versus that in developing countries is completely different. We do not have strong policies, strategies, and initiatives, both in the public and the industrial realms, and the connection between Industry and Research is still unconnected. This gap needs to be bridged—How? It is a big question for us. How to take up this challenge here in India would constitute great advice, if someone could take the lead and suggest something.

Shaver: I can perhaps give a quick look at what that "One Bin" project would mean. One of the things we have done is, we get paid to think about that material hierarchy from a UK infrastructure perspective. But we have taken that lens in terms of the potential decisions that you could make and looked at places such as the Philippines, where there is no centralized waste management infrastructure, and how this would change the potential decision-making or infrastructure. I think it is about building communities that understand the potential for it. But then I think it is really important to show the economic value of sustainability. This is true about Mike's work on green chemistry as well, right?

If it is presented only as an environmental issue, it is very difficult to motivate change across the sector. But if you show that economic growth is enabled alongside environmental sustainability, because you are derisking things, because compliance costs are less, and because systems are more efficient, that unlocks change. Tying those two narratives together has been really useful in our work with the UK government and with some international plastics trade bodies. I would be happy to follow up in more detail on that from a plastics front, but that is my initial advice.

Stenstrom: Yes. Well, we are at the end of our time. Did you want to speak? Well, we are at the end of our time. It is 3:01 my time in Los Angeles. So, I think we will close the session now. But I think we had a very good session, and I want to compliment the organizers of the conference in selecting very fine speakers and very fine commentators. I am really impressed with the knowledge that was displayed here tonight. So, we will stop now. I will give it back to the conference organizers. But this is good night from a very happy Prof. Stenstrom at 3 a.m. in the morning in Los Angeles, California. ◊

ICUS XXVIII

SESSION 3

Engaging the Public in Tackling Environmental Concerns

About Session 3

Engaging the Public in Tackling Environmental Concerns

The resolution of environmental problems requires the involvement and cooperation of all people, based on recognizing the value of the natural environment and understanding the outcome of our relationship with nature. This session examines how the public can be awakened and engaged using two approaches: policy making and education.

The first presentation calls attention to the problem of climate change, and the need for nations to implement policies that lead to net-zero emissions of greenhouse gases–that is, to achieve a "decarbonized" society. This presentation points to the commitments made by most nations through the Glasgow Pact, at the Twenty-Sixth Conference of the Parties (COP26) to the United Nations Framework Convention on Climate Change (UNFCCC), held in Glasgow, UK, in November 2021.

The second topic deals with education of the public at all levels, from formal schools to informal community settings. Education needs to not only provide the information but also promote pro-environmental attitudes and related skills. Thus, it needs to help people understand the issues and motivate them to adopt behaviors that protect and restore the natural environment. The challenge is to find ways to implement these ideas in a variety of social contexts. ◊

About the Chair, Suh-Yong Chung, JSD

- Director, Center for Global Climate and Marine Governance (CGCMG), affiliated with Global Research Institute, Korea University, Korea.
- Director, Center for Climate and Sustainable Development Law and Policy (CSDLAP), Korea.
- Professor, Division of International Studies, Korea University, Korea.
- The CGCMG focuses on approaches to implement the Paris Agreement, including forest capacity building in Africa, environmental ocean governance in Northeast Asia, low carbon technologies, environmental education, international cooperation and the carbon market, and research of the Arctic.
- JSD, Stanford University, USA.

Session 3 | Presentation One

Policy Challenges to Achieve a Net-Zero Society by 2050: A Perspective from Japan

Dr. Kazuo Matsushita

Professor Emeritus, Graduate School of Global Environmental Studies,
Kyoto University, Kyoto, Japan;
Senior Fellow, Institute of Global Environmental Strategies

COVID-19 and climate change are two urgent issues on which the international community has to work together in order to ensure the survival of humankind. Both problems are deeply related to economic globalization and urban concentration. We need to make an early transition to an economic society that is less prone to pandemics and that can avoid climate crises. The implementation of a "green recovery"—in other words, measures to restore the economy after the COVID-19 recession—should be an opportunity to realize a decarbonized society.

Green recovery from COVID-19

Conventional economic recovery metrics—such as growth in fossil fuel-intensive industries, expansion of construction projects, and so on—may augur short-term economic recovery but cannot produce long-term structural change to realize a decarbonized society. On the other hand, "green recovery measures" or the "Build Back Better" approach, which include new lifestyles and work styles, such as low-carbon employment, renewable energy, and telework, are expected to contribute to the transition to a decarbonized society and the realization of the United Nations'

About the Speaker

- Professor Emeritus, Graduate School of Global Environmental Studies, Kyoto University, Japan.
- Senior Fellow, Institute of Global Environmental Strategies, Japan.
- President, International Academic Society for Asian Community (ISAC), Japan.
- Chairman, Japan Society for Gross National Happiness Studies, Japan.
- MA in Political Economy, Johns Hopkins University, USA.

Sustainable Development Goals.

Unrestricted economic globalization prioritizes trade, capital liberalization, and maximizing trade flows, but the global supply chain should be reexamined from the perspective of increasing the resilience of local communities, countries, and the world against threats such as pandemics and climate change that endanger the sustainability of the international community.

Outcomes of COP26

COP26, the Twenty-Sixth Conference of the Parties (COP) to the United Nations Framework Convention on Climate Change (UNFCCC), held in Glasgow, UK, in November 2021, produced significant results toward the transition to a decarbonized society. The Glasgow Climate Pact (see https://unfccc.int/documents/310475), the outcome document adopted at COP26, yielded many achievements, and the following four points are particularly noteworthy:

1) The Paris Agreement's long-term target for temperature increase limitation was effectively strengthened from less than 2 °C to less than 1.5 °C.

2) For the first time, a policy to reduce coal-fired power generation as the biggest contributor to global warming was clearly stated in a COP decision.

3) The detailed rulebook of the Paris Agreement (guidelines for the implementation of Article 6, a market mechanism) was agreed upon, and the Paris Agreement was completed.

4) Funding for adaptation assistance to developing countries will double by 2025.

Then-British Prime Minister Boris Johnson, who chaired the conference, set *Coal* (phasing out coal-fired power generation), *Car* (conversion to electric vehicles), *Cash* (strengthening financial support for developing countries), and *Tree* (conservation of forests and expansion of afforestation) as the four key themes. On the basis of his initiatives, the following agreements were adopted during COP26 by the various countries and groups present:

- The Global Coal to Clean Power Transition Statement (see https://ukcop26.org/global-coal-to-clean-power-transition-statement/), a joint statement to end support for new coal-fired power generation, was announced on November 4, 2021. The agreement calls for developed countries to phase out unabated (without CO_2 emission reduction measures) coal-fired power generation in the 2030s at the latest, and for the world as a whole to phase it out in the 2040s. At the time of the announcement, 77 signatories, including 46 national governments (23 of which pledged for the first time, albeit with conditions, such as Indonesia, South Korea, Poland, Vietnam, and Chile), had joined in signing the statement. However, despite strong urging from the British government, Japan has not signed.

- Statement on International Public Support for the Transition to Clean Energy (see https://ukcop26.org/statement-on-international-public-support-for-the-clean-energy-transition), signed by 20 countries

- Public support for the fossil fuel energy sector ends in 2022

- Declaration on Accelerating the Transition to 100% Zero-Emission Passenger Cars and Vans (see https://www.gov.uk/government/publications/cop26-declaration-zero-emission-cars-and-vans/cop26-declaration-on-accelerating-the-transition-to-100-zero-emission-cars-and-vans)

- Glasgow Leaders' Declaration on Forests and Land Use (see https://ukcop26.org/glasgow-leaders-declaration-on-forests-and-land-use/)

After COP26, carbon neutrality has now become a global goal. By COP26, 136 countries, covering about 90% of global emissions, gross domestic product (GDP), and population, have signed on to the UN's net-zero initiative (see https://zerotracker.net/).

It was also made clear that closing the gap to achieve the 1.5 °C target is essential, since even achieving the current 2030 target fully would result in a 2.4 °C increase by the end of the twenty-first century. Therefore, strengthened action by 2030 is essential to close the gap with 1.5 °C (Climate Action Tracker 2021).

Carbon neutrality as a new national development strategy

Since COP26, it has become the de facto standard for major countries to aim for

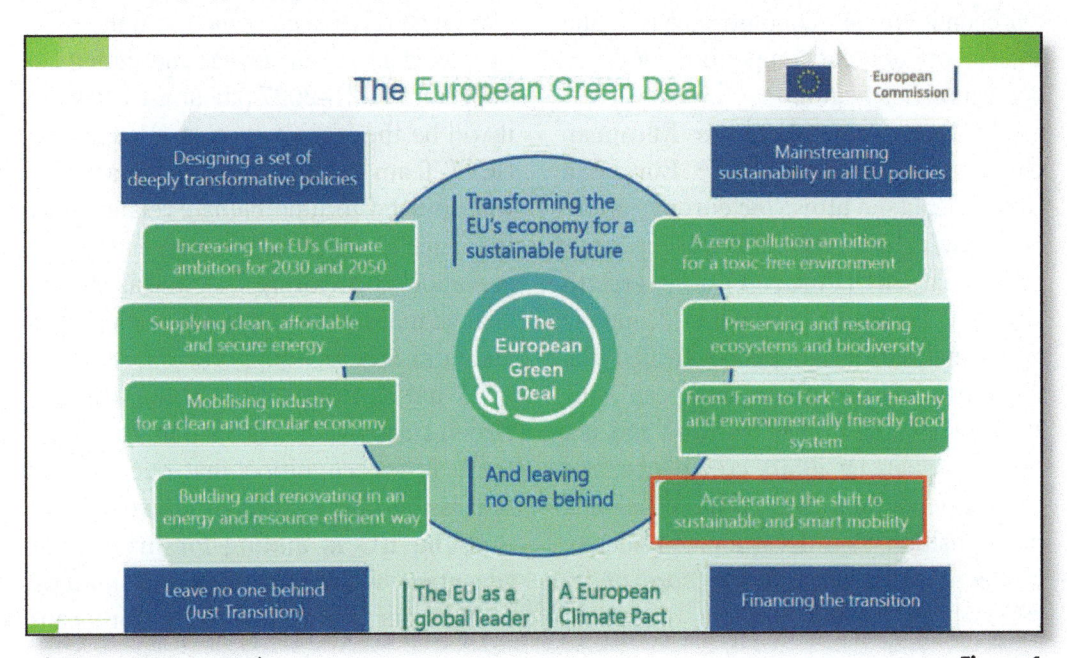

The European Green Deal

Figure 1

> **Next Generation EU (NGEU)**
>
> - Midterm budget after 2021 consists of Multiannual Financial Framework (MFF) and Recovery Fund (Next Generation EU).
>   NGEU (2021–2023) €750 billion, MFF (2021–2027) €1,074.3 billion
>
> - 30% of the total amount of €1.8 trillion will be spent on climate change measures: promotion of renewable energy and electric vehicles, R&D of hydrogen and fuel cells, promotion of energy conservation in buildings, etc.
> - Financing of the Recovery Fund: The European Commission will issue bonds on the financial markets. The bonds will be redeemed by the end of 2058 at the latest. The European Commission will expand the EU's own resources to cover the redemption of the Recovery Fund.
> - New sources of income will include the following
>   - Levy on single-use plastics (introduced in 2021)
>   - Carbon border-adjustment measures (tariffs on nonlow-carbon products from outside the EU expected to generate €10 billion per year) (proposed in June 2021)
> - Expansion of sectors subject to EU-ETS (emissions trading) (to include shipping and aviation sectors) (expected to generate €5–14 billion annually)
> - To be proposed by June 2024:
>   - Digital taxation
>   - Financial transaction tax
>   - New common consolidated corporate tax base, etc.

Figure 2

net-zero emissions of greenhouse gases (a decarbonized society) in their national development strategies. For countries and corporations, decarbonization has become a precondition for economic survival, and a global era of great competition for decarbonization has begun.

In December 2019, the European Union (EU) announced the European Green Deal (see https://ec.europa.eu/info/strategy/priorities-2019-2024/european-green-deal_en) (**Figure 1**) as a growth strategy, aiming to harmonize economic, production, and consumption activities with the Earth, reduce greenhouse gas emissions (a 55% reduction by 2030 and net-zero emissions by 2050), create jobs, and promote innovation. The EU has positioned the transition to a decarbonized society (net-zero emissions) as the core of its regional development strategy.

In July 2020, the EU agreed to establish the Next Generation EU Recovery Fund (**Figure 2**), which will issue bonds to raise €750 billion in addition to its regular budget. When combined with the EU's proposed next seven-year medium-term budget for 2021–2027 (about €1.7 trillion), it will be the largest ever, at €1.8 trillion (ICCJ Team 2021). Of this, at least 30% will be used to fund climate-related activities such as renewable energy, energy conservation, and hydrogen, as well as climate change measures such as electric vehicles and infrastructure support.

What is noteworthy about the EU Green Deal discussions is that, first of all, there is a recognition that decarbonization is the only pathway to develop, and that the EU is aiming for first-mover advantage by investing in decarbonization.

In order to achieve this, the Europeans

have drawn a concrete picture of what industry will look like in the decarbonization era and are discussing the path to get there and the policy measures that will facilitate the transition. By promoting the European Green Deal, the EU's standards and rules will eventually become international standards. Specifically, by establishing the EU taxonomy (see https://finance.ec.europa.eu/sustainable-finance/tools-and-standards/eu-taxonomy-sustainable-activities_en), the Europeans have introduced a common global definition of environmental, social, and governance investment and EU criteria for green investment. In addition, carbon border-adjustment measures will impose tariffs on non-low-carbon products from outside the region, thereby forcing environmental measures to affect nations outside the region.

In the US, under President Joseph Biden, the country returned to the Paris Agreement and is gradually realizing its campaign pledges on climate change, including 1) net-zero greenhouse gas emissions from the entire economy by 2050, 2) a US$2 trillion investment in sustainable infrastructure and clean energy over four years, 3) reinforcement of greenhouse gas emission regulations and incentives, 4) zero greenhouse gas emissions from the power generation sector by 2035, and 5) promoting environmental justice.

However, developments in Congress are hindering the implementation of President Biden's climate change commitments. Opposition from Republicans and a pro-fossil fuel member of the Democratic Party has led to the elimination or weakening of climate change measures in the president's H.R. 3684 (the Infrastructure Investment and Jobs Act; see https://www.congress.gov/bill/117th-congress/house-bill/3684/text) and H.R. 5376 (the Inflation Reduction Act of 2022; see https://www.congress.gov/bill/117th-congress/house-bill/5376/text).

The infrastructure investment bill will provide a total of US$1 trillion over five years, about US$280 billion for transportation infrastructure such as roads, bridges, and electric vehicle charging facilities, and about US$270 billion for non-transportation infrastructures such as water supply and high-speed communication networks. This bill was passed by Congress on November 6, 2021, despite difficulties in reconciliation among Democrats.

China also announced emissions reduction targets. President Xi Jinping promulgated a plan to reduce CO_2 emissions by 2030 and achieve net-zero CO_2 emissions by 2060 at the U.N. General Assembly on September 22, 2020. China is the world's largest emitter of CO_2 (28% of the global total), and this announcement was greeted with surprise by the world.

In previous international negotiations, China routinely called attention to the historical emissions responsibility of developed countries, and as a developing country has refrained from setting a total emissions reduction target. This change in policy was significant. It may have come about due to domestic factors such as a political judgment that efforts toward carbon neutrality will lead to high-quality economic development. In addition, the turnabout may have been impelled by scientific research reports supporting such a transition—for

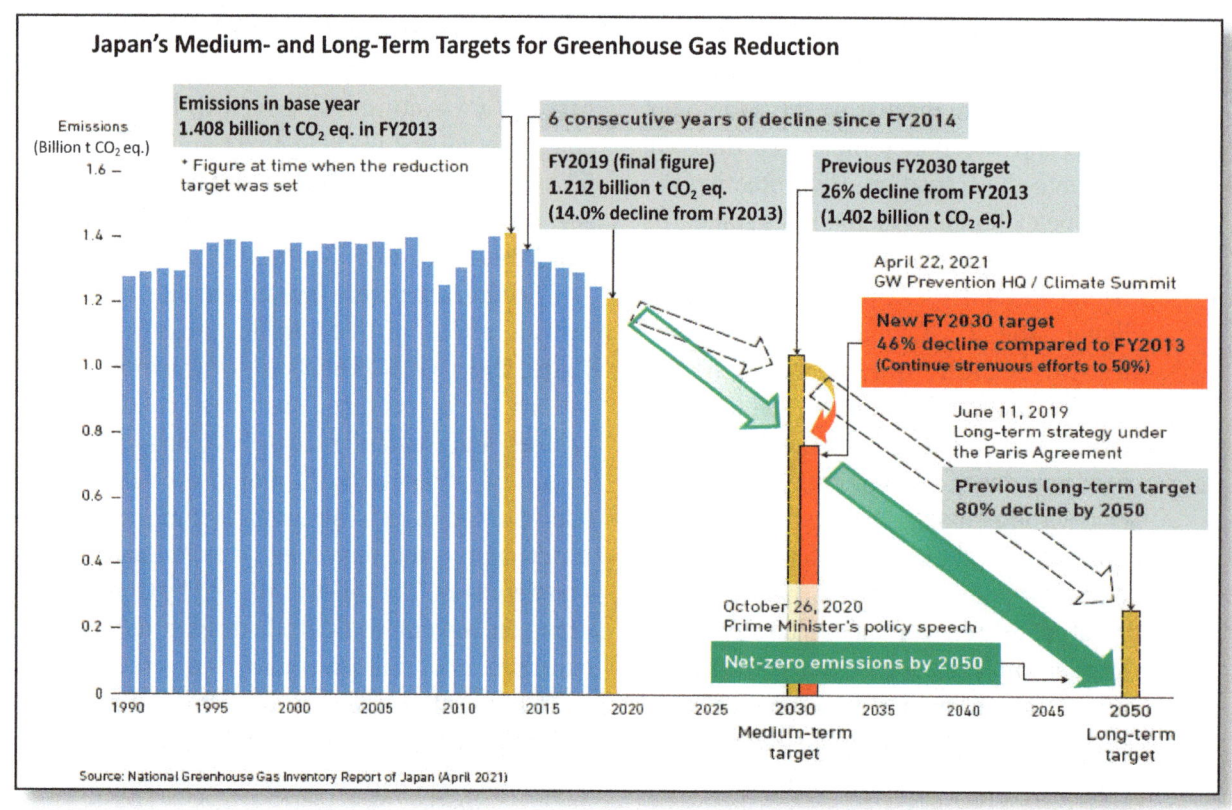

Figure 3

Japan's Medium- and Long-Term Targets for Greenhouse Gas Reduction

Source: National Greenhouse Gas Inventory Report of Japan (April 2021)

example, the scenario envisioned by Kyoto Seika University of zero CO_2 emissions from the power sector in 50 years.

Challenges for Japan: From green recovery to a net-zero society

In Japan, former Prime Minister Yoshihide Suga declared in an address to parliament (Suga 2020) in October 2020, "We aim to reduce greenhouse gas emissions to net zero (carbon neutral) by 2050 (**Figure 3**) and to realize a decarbonized society." He went on to say, "Coping with global warming is no longer a constraint on economic growth. We need to change our mindset and realize that proactive measures to combat global warming will lead to changes in industrial structure and economic society, which in turn will lead to significant growth." In addition to technological innovation, he called for regulatory reform and the acceleration of green investment, and he mentioned the environmental problems arising from the digitization of society.

Prime Minister Fumio Kishida spoke at COP26 on November 2, 2021 (Kishida 2021). In his address, he declared Japan's intention to increase contributions to international climate funds, which was welcomed. However, there was no clear reference to the 1.5 °C target of the Paris Agreement or to a statement of political will to strengthen the 2030 target. There was no mention of phasing out domestic coal-fired power plants and suspension of support for overseas coal-fired power generation projects.

He laid out a plan to develop technologies to use hydrogen and ammonia for power generation both at home and abroad. However, the technologies to use hydrogen and ammonia as fuels have not yet been established on a marketable basis, and, according to the plan of the government, full-scale commercialization is not expected to be feasible until the 2040s or later. Moreover, we cannot presently say with confidence that we will be able to decarbonize if we gasify coal to produce hydrogen (brown hydrogen) or use thermal power to produce ammonia. These are, in effect, measures to extend the life of coal-fired power plants, and it is unreasonable to claim that such coal-fired plants have been "abated" (Japan Beyond Coal 2022).

What is expected of Japan in the wake of COP26?

In the wake of COP26, the following measures are required to be adopted by the government of Japan to attain the goals of the Paris Agreement and to achieve a net-zero society by 2050:

- Clearly set a target to limit temperature increase by 1.5 °C.

- Strengthen the 2030 target and introduce a five-year review system.

- Strengthen policy measures to raise the target, such as:

    o Adopt a phase-out policy for coal-fired power plants.

    o Transform fiscal measures from "innovation" to "just transition."

    o Introduce full-fledged carbon pricing.

- Enhance energy conservation in residential construction.

- Increase support for developing countries.

- Make the decision-making process more participatory and transparent.

Phasing out coal-fired power generation, the largest source of emissions

In the run-up to COP26, British Prime Minister Johnson, the leader of the presiding country, strongly urged that coal-fired power generation be "phased out by 2030 for developed countries and by 2040 for developing countries."

On November 4, 2021, during COP26, 46 countries and regions, including the United Kingdom, Germany, France, and the EU, signed a statement that included the following language: "Developed countries will phase out coal-fired power generation in the 2030s, and other countries will phase out coal-fired power generation in the 2040s, or stop new construction" (see https://ukcop26.org/global-coal-to-clean-power-transition-statement/). Japan, however, did not sign this statement and is now the only G7 country that has not specified a time frame for eliminating coal-fired power generation (see **Table 1**).

Country	Coal-Fired Power Policy
Japan	Coal-fired power source ratio is planned to be 19% in Japan in FY2030, New construction planned
United States	Decarbonize the power sector by 2035
United Kingdom	Eliminate domestically by 2024; require developed countries to phase out by 2030 and other countries by 2040
France	Abolish by 2022
Germany	New government: (ideally) abolish by 2030 (previously, abolish by 2038)
Canada	Abolish by 2030
China	Announced suspension of export support to foreign countries in September 2021; numerous plans for new construction in the country

Table 1 Positions of major countries regarding coal-fired thermal power (prepared by the author from various sources)

With regard to coal-fired power generation, the Glasgow Climate Pact (see https://unfccc.int/documents/310475) pledges "to accelerate efforts to phase down unabated coal-fired power generation and to phase out inefficient fossil fuel subsidies." Although toned down by India's proposal, the global trend to eliminate coal-fired power generation is unstoppable.

However, the Japanese government's Sixth Strategic Energy Plan, announced in October 2021 (see https://www.meti.go.jp/english/press/2021/1022_002.html), still expects coal-fired power to account for about 19% of the power supply in the fiscal year of 2030. It is imperative for Japan to decide to phase out coal-fired power generation by 2030 and to develop phased transition measures to achieve this goal.

Introduction of a full-scale carbon pricing system

Now that the rules of Article 6 of the Paris Agreement (see https://unfccc.int/files/meetings/paris_nov_2015/application/pdf/paris_agreement_english_.pdf) have been decided, it is expected that a huge global carbon trading market will be launched. In addition, the EU and the US are beginning to discuss border carbon-adjustment mechanisms in earnest. For these reasons, it is desirable to introduce full-scale carbon pricing in Japan as soon as possible.

Carbon pricing is a policy that aims to reduce emissions by putting a price on carbon dioxide, and there are two ways to implement it: carbon taxes and emissions trading. Such strategies create a powerful price signal to guide economic actors toward a low-carbon society. A gradual increase in the price of carbon to achieve the goal of a decarbonized society will spur technological innovation and the development of low-carbon infrastructure, accelerating the transition to zero- or low-carbon goods and services.

Government revenues from carbon

pricing may be used to reduce social security costs, provide income benefits to low-income groups, and invest in the energy transition. In parallel, the transition to energy conservation and renewable energy will be further facilitated by ending fossil fuel incentives such as subsidies and tax breaks. The introduction of carbon pricing will promote the simultaneous achievement of higher carbon productivity and higher profitability, encouraging a shift from carbon-intensive and low-profit business areas to low-carbon but highly profitable businesses.

In Japan, a global warming tax (carbon tax) was introduced in 2012, but it has not been effective in curbing CO_2 emissions because the tax amount per metric ton of CO_2 emissions is just ¥289, which is significantly lower than in other countries that have introduced carbon taxes. Nations that have adopted effective carbon taxes since the 1990s have seen their CO_2 emissions decline while their GDPs have grown—a process known as *decoupling*. In Scandinavian countries, for example, setting a high carbon-tax rate has encouraged the spread of products that emit less CO_2 and the development of energy-saving technologies, leading to new economic development. In Japan, on the other hand, CO_2 emissions have increased while GDP has remained flat.

The time to achieve the 2030 target and transition to a decarbonized society by 2050 is running out. Prompt introduction of full-scale carbon pricing under an appropriate institutional design is highly desirable.

Conclusion

The climate crisis is a crisis of human rights in the sense that it threatens the basis of people's survival, and in particular, it is a crisis of children's rights that deprives future generations of the possibility of development. Therefore, achieving carbon neutrality by 2050 is a way of fulfilling our responsibility to future generations.

In order to achieve the 1.5 °C target of the Paris Agreement, we need to reduce greenhouse gas emissions by 45% globally by 2030 and achieve net-zero (a decarbonized society) by 2050. The transition to a decarbonized society has already begun. However, we have little time left.

In order to attain a net-zero society by 2050, we need a backcasting approach. That is, we need to envision a decarbonized 2050 society and then develop an overall strategy, based on sound scientific evidence, to achieve it. From now until 2030, we must maximize the use of the best available technologies and implement institutional and financial reforms. After that, until 2050, our goal should be to encourage innovation in order to accelerate the transition to a new society.

In the post-COP26 world, zero emissions has become the standard as a new national development strategy, and decarbonization has become a precondition for economic survival. The era of great decarbonization competition has already begun.

Japan has long lagged behind in setting ambitious goals for decarbonization, has postponed the introduction of economic stimulus measures such as full-scale carbon pricing, and has continued to rely excessively on coal-fired power plants. As a result, it is a regret to say that the transition to a decarbonized society has lagged far behind.

The world's efforts to transition to a decarbonized society are moving quite rapidly. It is imperative for Japan to bring about a society that is decarbonized, symbiotic with nature, circular, and regionally self-reliant, where people can live humanely and with dignity. ◊

References

Climate Action Tracker. November 2021. "Temperatures: Addressing Global Warming." Climate Action Tracker. Accessed on September 13, 2022.
https://climateactiontracker.org/global/temperatures/

ICCJ Team. November 30, 2021. "NEXT GENERATION EU Fund:
The European Plan to Recover the Economy."
Italian Chamber of Commerce in Japan.
Accessed on September 13, 2022.
https://iccj.or.jp/next-genereation-eu-fund-the-european-plan-to-recover-the-economy/

Japan Beyond Coal. January 7, 2022. "With No Coal Exit Policy, Japan Falls Short at COP26." Japan Beyond Coal. Accessed on September 13, 2022.
https://beyond-coal.jp/en/news/cop26-deviate-japan/

Kishida, Fumio. November 2, 2021. "COP26 World Leaders Summit Statement by Prime Minister Kishida Fumio." Provisional translation. Prime Minister of Japan and His Cabinet. Accessed on September 13, 2022.
https://japan.kantei.go.jp/100_kishida/statement/202111/_00002.html

Suga, Yoshihide. October 28, 2020. "Policy Speech by the Prime Minister to the 203rd Session of the Diet." Provisional translation. Prime Minister of Japan and His Cabinet. Accessed on September 13, 2022.
https://japan.kantei.go.jp/99_suga/statement/202010/_00006.html

Session 3 | First Commentary
on the Presentation by Dr. Kazuo Matsushita

Policy Challenges to Achieve a Net-Zero Society by 2050: A Perspective from Japan

Professor Wil Burns
Environmental Policy & Culture Program, Northwestern University, Illinois, USA

Dr. Kazuo Matsushita's presentation, "Policy Challenges to Achieve a Net-Zero Society by 2050: A Perspective from Japan," provides an excellent overview of both the prospects and challenges of achieving what many believe is a critical objective—reaching net-zero emissions by 2050. In my role as a respondent, I would like to briefly discuss Dr. Matsushita's observations and seek to parse out their implications for the future.

The Green Recovery

Dr. Matsushita mentioned at the outset of his presentation the twin crises of COVID-19 and climate change. One of the few silver linings of the pandemic was that global CO_2 emissions dropped by about 5.2% in 2020 due to reductions in economic activity. However, as the world economy recovered in 2021, it provided a sobering reminder of how far we have to go to structurally decarbonize the global economy, which was the focus of Dr. Matsushita's presentation. Global energy-related carbon dioxide emissions rose by 6% in 2021 to 36.3 billion t, their highest level ever, eclipsing the decline during the height of COVID-19 in 2020.

About the Speaker

- Visiting Professor, Environmental Policy and Culture Program, Northwestern University, Illinois, USA.
- Founding Co-Executive Director, Institute for Carbon Removal Law and Policy, American University, USA.
- Former President, Association of Environmental Studies & Sciences.
- Research Fellow, Center for Science, Technology, Medicine & Society, University of California, Berkeley, USA.
- Founding Editor in Chief, *Case Studies in the Environment*, University of California Press, USA.
- PhD in International Environmental Law, University of Wales, UK.

The increase in global CO_2 emissions of over 2 billion tons in 2021 was the largest in history in absolute terms.

The green recovery measures that Dr. Matsushita emphasized are critical for meeting the objectives of the Paris Agreement but are a relatively small component of total COVID-19 designated spending, estimated at only 21%. Moreover, significant portions of recovery funds are being allocated to measures with environmentally negative or mixed impacts, with substantial funding for projects that may create substantial amounts of emissions lock-in. It is also sobering to note that investments in sustainable energy projects would still represent only 40% of the projected investment needs to meet the objectives of the Paris Agreement. This is according to the International Energy Agency's Sustainable Recovery Plan, which is aligned with a pathway toward reaching net-zero emissions by 2050 globally.

As Dr. Matsushita also observed, a number of important emitting nations are taking measures to meet the objectives growing out of the agreements made at COP26 on climate change and the net-zero commitments there. However, if one looks at trends in the world's two largest emitters, one must be skeptical—first of all in terms of China. As Dr. Matsushita indicated, China is the world's greenhouse gas (GHG) behemoth. It accounts for approximately 28% of all CO_2 emissions.

To put this in perspective, China's GHG emissions are now greater than those of all developing countries combined, as well as greater than all developed countries' combined emissions.

In 2021, China's CO_2 emissions rose to 11.9 billion tons, accounting for 33% of the global total. The country's emissions are more than triple the 1990s levels and are up 25% in the past decade. The emissions increases in 2020 and 2021 in China more than offset the aggregate decline in the rest of the world over the same period. On an annual per capita basis, CO_2 emissions in advanced economies have fallen to 8.2 tons on average and thus are now below the average of 8.4 tons per capita in China, although wide differences remain among advanced economies.

Moreover, China does not appear to be taking the measures that could result in it meeting its own pronounced objectives of achieving a peaking of emissions in 2030 and achieving net-zero emissions by 2050. Coal is responsible for approximately 70% of China's energy consumption in the past decade and produces about 75% of the nation's CO_2 emissions. President Xi Jinping has vowed to limit increases in coal-fired power plant emissions in the next five years, and then begin to reduce them. However, it is not clear if this is little more than puffery at this point. Chinese officials have also emphasized that economic growth, which is still largely dependent on coal power, remains a priority, and the nation is still engaged in massive construction of coal-fired power plants. More than 1,000 coal plants are in operation, and almost 240 are planned or already under construction. Thus, it strains credulity that China will allow for massive stranded assets in a couple of years in terms of new coal plant infrastructure.

In the United States, greenhouse gas

emissions rose a whopping 6.2% in 2021 above 2020 levels. This jump in emissions suggests that the US is not on track to meet its recently announced commitment under the Paris Agreement to cut 2005 emission levels by 50–52% by 2030. To put in perspective how imposing this goal is, the Rhodium Group recently projected that the US has to reduce its annual emissions by about 5% each year to meet the goal of net zero by 2050. However, reductions of this size are unprecedented in US history, other than during the COVID-19 shutdown.

I also concur with Dr. Matsushita that Japan, as an advanced economy with the fifth-largest GHG emissions in the world, is dragging its feet in terms of committing to a reduction path that comports with the Paris Agreement. Unfortunately, if larger economies, such as China and the US are not willing to take the requisite measures to decarbonize, other major economies, such as Japan, may be chary to take measures that could place it at a competitive disadvantage.

As Dr. Matsushita also points out in his presentation, phase-out of coal-fired power plants, the largest source of CO_2 emissions in the world, is critical to meeting the global temperature objectives of the Paris Agreement. However, the world's latest crisis, the war in Ukraine, may upend this. If the war precipitates reductions in natural gas supplies in Europe and elsewhere, which now seems inevitable, we may see another spike in coal utilization, slowing down the world's energy transition. On the other hand, one of the few silver linings of the war may be an increased commitment by many nations to accelerating the transition to renewable energy sources and energy efficiency. However, until the world massively reduces its dependence on fossil fuels, events of this nature may continue to prove economically wrenching, and may further delay the path to net-zero emissions.

The Way Forward

Dr. Matsushita also does an excellent job of explaining why carbon pricing mechanisms, such as carbon taxes or emissions trading systems, are important in terms of sending the kind of price signals that can spur fuel switching and technological innovation. However, as he also points out in his presentation, some of these domestic systems have failed to effectuate these goals because they set carbon taxes too low or emissions caps too high, which is largely related to politics.

This remains an extremely difficult political question, even in countries such as the United States that ostensibly extol the virtues of the free market, and thus should embrace the capturing of externalities associated with energy production through pricing mechanisms, yet do not because of the political influence of vested interests. I would argue in favor of an international pricing mechanism to try to break the Gordian knot of domestic politics, although it is hard to envision the global community embracing this approach anytime soon.

One final observation I would make is that meeting the objectives of the Paris Agreement now requires opening up yet another significant front in the war on cli-

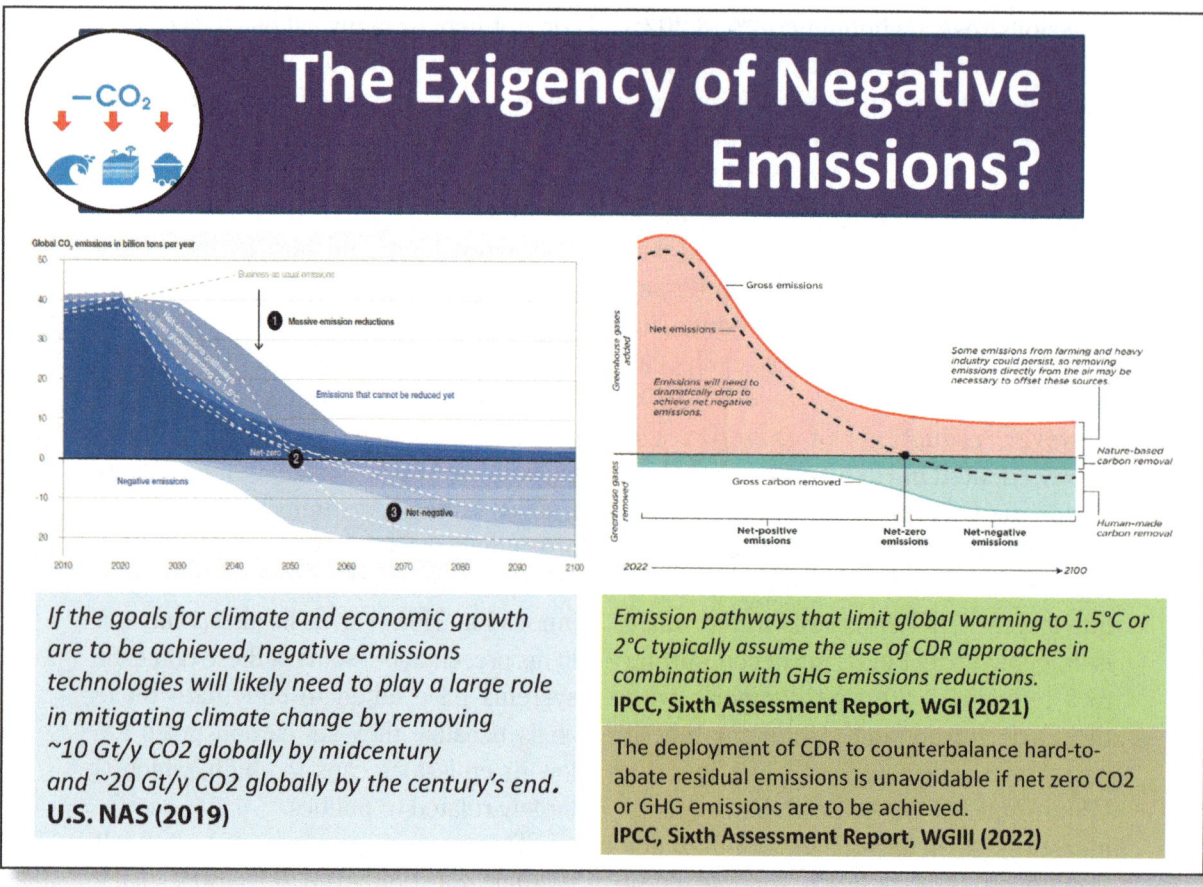

Figure 1

mate change: large-scale deployment of carbon dioxide removal (CDR) options. This was emphasized in both the Fifth and Sixth Assessment Reports of the Intergovernmental Panel on Climate Change (IPCC). The IPCC has provided a compelling need for both decarbonization and large-scale CDR at this point—CDR as high as 15–20 Gt/year of removal by the end of the century (**Figure 1**). These approaches include afforestation and reforestation, direct air capture, bioenergy with carbon capture and sequestration, enhanced mineral weathering, biochar, and a number of ocean-based options, such as ocean alkalinity enhancement and ocean iron fertilization (**Figure 2**). However, many of these approaches are nascent, and many could pose risks, especially at large scales of deployment. The world community needs to mobilize quickly to deploy these approaches if they are to meaningfully contribute to efforts to avoid passing critical climatic thresholds. However, it is critical to ensure that we minimize environmental risks and potential social inequities.

I also concur with Dr. Matsushita that climate change is an existential threat to future generations as well as a fundamental human rights issue. Those countries

How Hypothetical Technologies Shape Climate Scenarios

Most climate model scenarios rely on carbon dioxide removal (CDR) technologies to limit future temperature rises. Reliance on these technologies in models is problematic because they remain untested at the required scales.

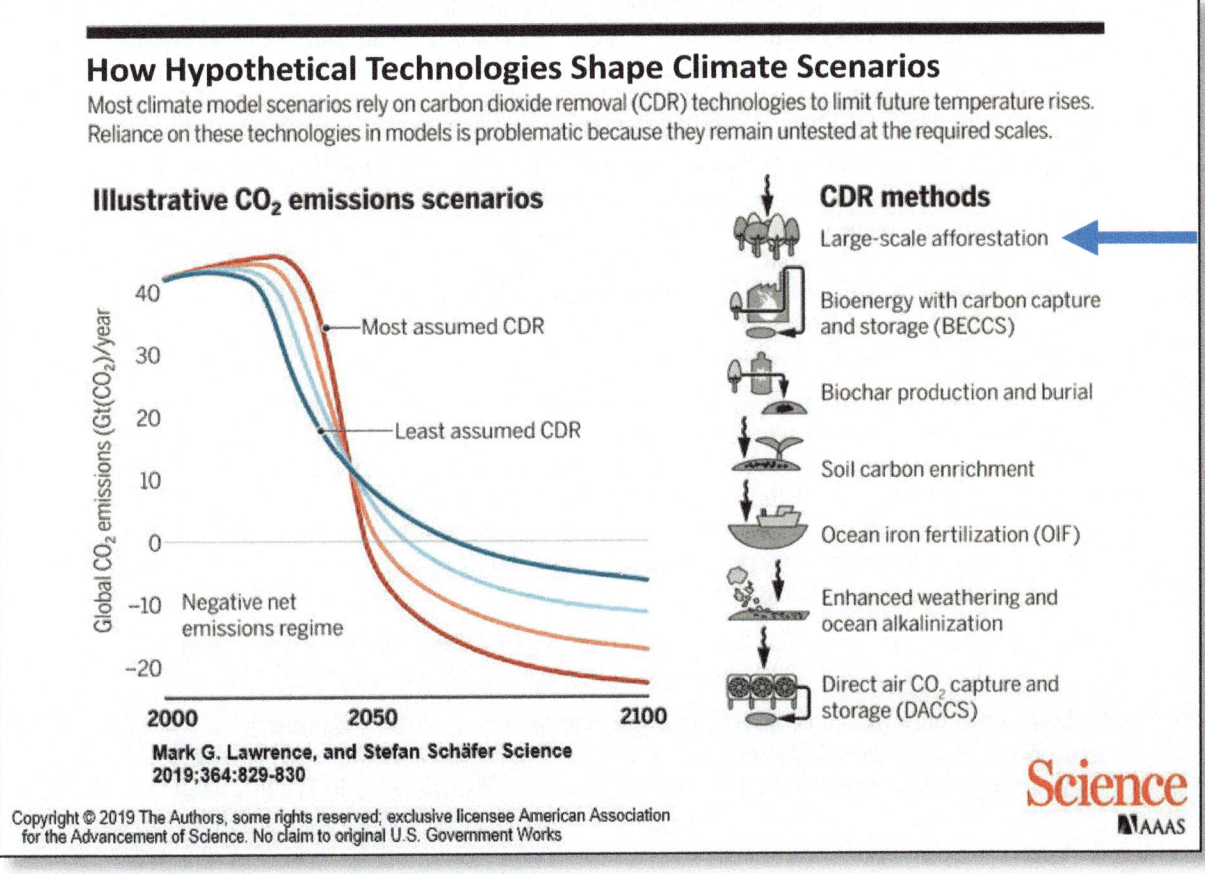

Illustrative CO₂ emissions scenarios

- Most assumed CDR
- Least assumed CDR
- Negative net emissions regime

CDR methods
- Large-scale afforestation
- Bioenergy with carbon capture and storage (BECCS)
- Biochar production and burial
- Soil carbon enrichment
- Ocean iron fertilization (OIF)
- Enhanced weathering and ocean alkalinization
- Direct air CO₂ capture and storage (DACCS)

Mark G. Lawrence, and Stefan Schäfer Science 2019;364:829-830

Copyright © 2019 The Authors, some rights reserved; exclusive licensee American Association for the Advancement of Science. No claim to original U.S. Government Works

Figure 2

most responsible for the current crisis bear heavy responsibility to substantially strengthen their emissions reduction commitments. It is my hope that the increasing focus on this issue by United Nations human rights institutions will help impel the world community to take the challenge of climate change more seriously, for the sake of this generation and future generations. ◊

References

Arias, P.A., N. Bellouin, E. Coppola, R.G. Jones, G. Krinner, J. Marotzke, V. Naik, M.D. Palmer, G.-K. Plattner, J. Rogelj, M. Rojas, J. Sillmann, T. Storelvmo, P.W. Thorne, B. Trewin, K. Achuta Rao, B. Adhikary, R.P. Allan, K. Armour, G. Bala, R. Barimalala, S. Berger, J.G. Canadell, C. Cassou, A. Cherchi, W. Collins, W.D. Collins, S.L. Connors, S. Corti, F. Cruz, F.J. Dentener, C. Dereczynski, A. Di Luca, A. Diongue Niang, F.J. Doblas-Reyes, A. Dosio, H. Douville, F. Engelbrecht, V. Eyring, E. Fischer, P. Forster, B. Fox-Kemper, J.S. Fuglestvedt, J.C. Fyfe, N.P. Gillett, L. Goldfarb, I. Gorodetskaya, J.M. Gutierrez, R. Hamdi, E. Hawkins, H.T. Hewitt, P. Hope, A.S. Islam, C. Jones, D.S. Kaufman, R.E. Kopp, Y. Kosaka, J. Kossin, S. Krakovska,

J.-Y. Lee, J. Li, T. Mauritsen, T.K. Maycock, M. Meinshausen, S.-K. Min, P.M.S. Monteiro, T. Ngo-Duc, F. Otto, I. Pinto, A. Pirani, K. Raghavan, R. Ranasinghe, A.C. Ruane, L. Ruiz, J.-B. Sallée, B.H. Samset, S. Sathyendranath, S.I. Seneviratne, A.A. Sörensson, S. Szopa, I. Takayabu, A.-M. Tréguier, B. van den Hurk, R. Vautard, K. von Schuckmann, S. Zaehle, X. Zhang, and K. Zickfeld. "Technical Summary." In *Climate Change 2021: The Physical Science Basis. Contribution of Working Group I to the Sixth Assessment Report of the Intergovernmental Panel on Climate Change*, edited by V. Masson-Delmotte, V., P. Zhai, A. Pirani, S.L. Connors, C. Péan, S. Berger, N. Caud, Y. Chen, L. Goldfarb, M.I. Gomis, M. Huang, K. Leitzell, E. Lonnoy, J.B.R. Matthews, T.K. Maycock, T. Waterfield, O. Yelekçi, R. Yu, and B. Zhou. Cambridge, UK, and New York: Cambridge University Press, 33–144. doi:10.1017/9781009157896.002.

IPCC (Intergovernmental Panel on Climate Change). 2022. "Summary for Policymakers." In *Climate Change 2022: Mitigation of Climate Change. Contribution of Working Group III to the Sixth Assessment Report of the Intergovernmental Panel on Climate Change*, edited by P.R. Shukla, J. Skea, R. Slade, A. Al Khourdajie, R. van Diemen, D. McCollum, M. Pathak, S. Some, P. Vyas, R. Fradera, M. Belkacemi, A. Hasija, G. Lisboa, S. Luz, and J. Malley. Cambridge, UK, and New York: Cambridge University Press. doi: 10.1017/9781009157926.001.

Lawrence, M.G., and S. Schäfer. 2019. "Promises and Perils of the Paris Agreement." *Science* 364 (6443): 829–830. https://doi.org/10.1126/science.aaw4602

National Academies of Sciences, Engineering, and Medicine. 2019. "Summary." In *Negative Emissions Technologies and Reliable Sequestration: A Research Agenda*, 1–22. Washington, DC: The National Academies Press. https://doi.org/10.17226/25259

Session 3 | Second Commentary
on the Presentation by Dr. Kazuo Matsushita

Appraising Topics that Aim to Restore the Environment

Professor Marilyn A. Brown
Sustainable Systems, School of Public Policy,
Georgia Institute of Technology, Atlanta, USA

Thank you very much for the opportunity to speak at the Twenty-Eighth International Conference on the Unity of the Sciences and in particular to reflect on the comments of Dr. Kazuo Matsushita regarding policy challenges to achieve a net-zero society by 2050.

The doctor mentions moving to a net-zero society through the help of a green recovery, which is very reflective of the US focus on "building back better" from the COVID-19 devastation—in particular, the focus on phasing out coal and pricing carbon. Both of those are high priorities in the United States. We are experiencing a slow response to the challenge of climate change. Avoiding costly climate change necessitates action at every political level and across all sectors of the economy. The climate countdown has begun, but progress is slow. One reason is that we lack local road maps for how to create solutions that can be enthusiastically supported at all levels of society.

I would like to organize my comments around the research of the Intergovernmental Panel on Climate Change (IPCC), an organization that I have worked on for two decades, and in particular, to talk about their latest two reports. One is on the science of climate change. This produced the starkest warning yet about the demise of our global climate regime. It helped last year's COP26 meeting

About the Speaker

- Regents and Brook Byers Professor of Sustainable Systems; Co-Director, Climate and Energy Policy Lab, Georgia Institute of Technology, USA.
- Former Director, Oak Ridge National Laboratory, USA.
- Member of the National Academy of Engineering and National Academy of Sciences.
- Recipient of the World Citizen Prize in Environmental Performance (2021) and corecipient of the 2007 Nobel Peace Prize as a coauthor of the Intergovernmental Panel on Climate Change Working Group III Assessment Report on Mitigation of Climate Change, Chapter 6.
- PhD in Geography, The Ohio State University, USA.

in Glasgow to sound an alarm that immediate action to mitigate emissions is needed.

The second IPCC report is one I would like to dig a little bit deeper into and reflect on. That was published about a month ago and is called *Climate Change 2022: Impacts, Adaptation, and Vulnerability*. It has been much anticipated, because it will inform negotiations at the next Conference of the Parties in Egypt. It has already trained a spotlight on climate justice, and some of my remarks from the IPCC report on impacts will deal with this topic.

These two reports offer four unambiguous takeaways.

The first one is the need to focus on here-and-now messaging. We need to talk about climate impacts in the present tense not the future tense and not discuss them as events that are taking place somewhere else, such as in the Global South or some other distant location. Impacts are being felt everywhere. This here-and-now tone is so important for making policies more impactful.

Second, there are hard limits to adaptation. This is a new message. Some climate impacts will be harmful no matter how they are addressed, and examples include coral reefs that we are losing, rain forests, coastal wetlands, and the polar ecosystem. Those are all ecosystem impacts. There are also human hard limits to adaptation. One example here in my home state of Georgia is the threat that heat and humidity can pose, making unprotected outdoor life impossible. That is a hard limit to climate change.

The third takeaway from the IPCC report regards the social and health dangers of inequality. Where there are great extremes of poverty and wealth, those who are economically disadvantaged are more vulnerable to the impacts of climate change. For example, if you need to have broad participation and a consensus around actions, that is most achievable in a society that is not extremely unequal.

Climate change actually intensifies inequality, making lower-income urban neighborhoods more vulnerable to changes in the climate. As one example, a community science field experiment in the city of Atlanta showed the uneven distribution of heat and humidity across the metropolis. Sampling took place at three different hours on one day last summer, and the result was evidence of economically depressed areas of the city—those with a paucity of trees and parks—suffering under heat that was 5–6°F greater than the temperature in more well-off neighborhoods. The poorer areas were noted to be in parts of the city with majority black populations, neighborhoods that had been historically redlined by banks and other lenders, tending to lock in the areas' poverty. Heat injustice is one of those hard limits and a consequence of economic and social inequality.

Fourth and finally, climate change is more than an environmental catastrophe. It is a threat multiplier for the military, it creates displacement and migration perils for nation-states, and, for some societies, it causes dangerous competition for food, energy, resources, and more. In some cases, climate change impacts all the dimensions of life worth living.

Hopefully, these challenges will motivate future generations of students and our scientists to consider climate implications when assessing policy priorities in whatever fields of specialization they are engaged in. Thank you very much. ◊

Session 3 | Presentation Two

Promoting Grassroots Action on Environmental Issues

Professor Bruce Johnson
Environmental Learning, University of Arizona, Tucson, USA;
International Program and Research Coordinator, The Institute for Earth Education

Making meaningful changes related to solving current and future environmental problems is multifaceted. It is not enough for focus to be on only one aspect, such as technology, policy, or personal or collective behavior. All must be addressed. There is a role for education in all of these—but up to this point, education has not stepped up.

While there have been and continue to be numerous attempts to bring the environmental crises of our day into formal education systems around the world, those that have been successful have been on small, often local scales. There has been very little in the way of a serious, comprehensive educational response to the environmental threats we face. Education, however, goes far beyond formal schooling. Education in informal contexts is in many cases more influential in areas such as science education and environmental learning. Additionally, education can inform us about ways to successfully encourage people to adopt, both individually and collectively, more environmentally friendly behaviors.

For people to make meaningful changes in their individual or collective actions, knowledge alone is not enough.

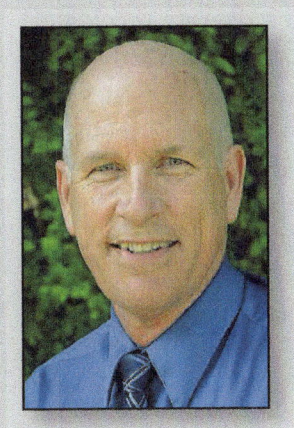

About the Speaker

- Professor of Environmental Learning and Science Education, Dean Emeritus, and former Paul L. Lindsey & Kathy J. Alexander Chair, College of Education, University of Arizona, USA.
- Researches the teaching and learning of ecological concepts; development of environmental values, attitudes, and actions; and Earth education curriculum development.
- International Program Coordinator, The Institute for Earth Education, USA.
- Member of the Editorial Boards of *Learning Environments Research* and *Envigogika*.
- PhD in Education Psychology, University of New Mexico, USA.

Promoting Grassroots Action on Environmental Issues

Relationships among Knowledge, Attitudes, and Behavior

[Diagram: Three circles — "Environmental system knowledge" with arrow to "Action-related knowledge"; both with arrows pointing down to "Effectiveness knowledge"]

Competence Model for Environmental Education
(Roczen, Kaiser, Bogner & Wilson, 2013)

THE UNIVERSITY OF ARIZONA
College of Education

The Institute for Earth Education

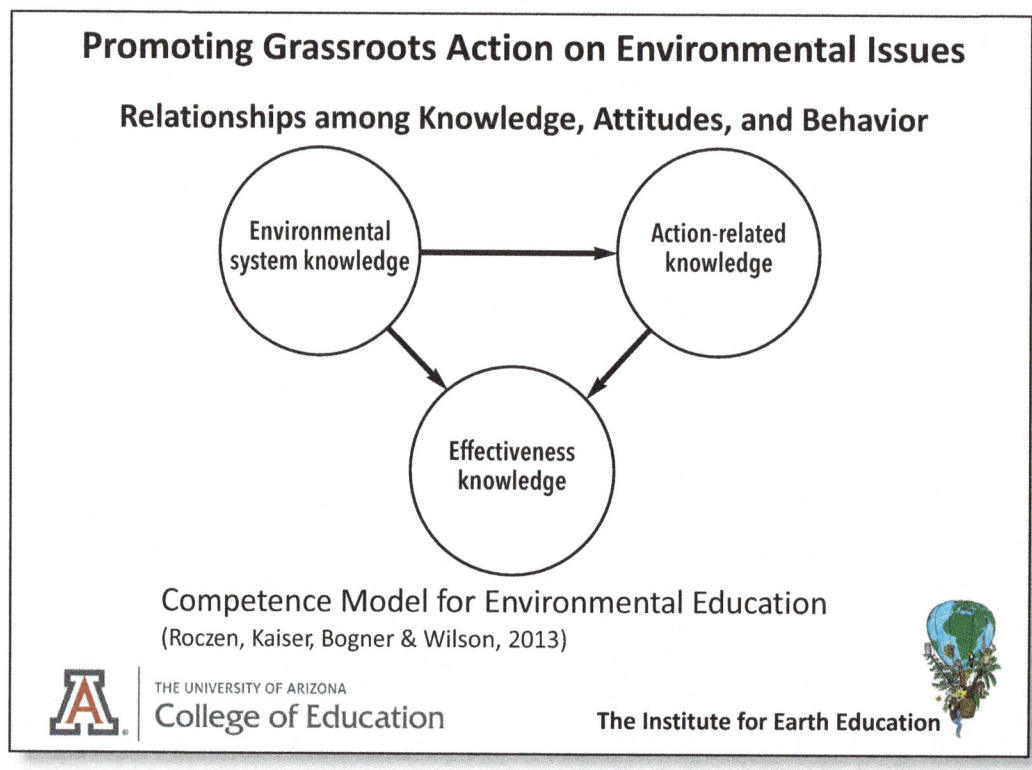

Figure 1 Knowledge model

Many other factors, including values, attitudes, and behavioral intentions, are critical. This paper examines the issue by looking at research in three areas: 1) the relationships among knowledge, attitudes, and environmental behavior; 2) psychological research on values and environmental identity; and 3) earth education programs designed to influence environmental action and behavior.

Relationships among knowledge, attitudes, and environmental behavior

The clearest paradigm of the relationships among knowledge, attitudes, and behavior is the Competence Model for Environmental Education (Roczen et al. 2013). The first part of the model differentiates between different types of knowledge—system, action related, and effectiveness (see **Figure 1**).

System knowledge is the dominant type of environmental knowledge that is taught and forms the basis for the other knowledge dimensions. This knowledge can include how ecological systems work and natural systems operate (ecological concepts) but can also embody information about environmental issues and impacts. Another way to think about this is as "knowing what."

Action-related knowledge is knowledge of behavioral options and possible courses of action, or "knowing how" to be more environmentally friendly. This

Promoting Grassroots Action on Environmental Issues

Attitudes toward Nature

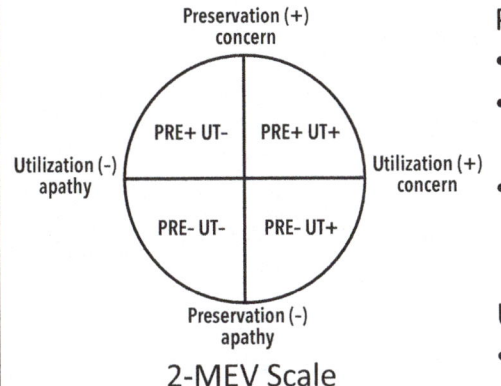

2-MEV Scale
(Bogner & Wiseman, 2006)

THE UNIVERSITY OF ARIZONA
College of Education

Preservation of Nature:
- care toward resources
- support of environmental causes
- enjoyment and appreciation of nature

Utilization of Nature:
- altering nature
- human dominance over nature

The Institute for Earth Education

Figure 2

can include individual actions that can be taken (reducing electricity use, recycling, and so on) or group actions, such as working to change policies. Action-related knowledge is not taught as frequently as systems knowledge.

Effectiveness knowledge is about the relative gain or benefit that is associated with a particular behavior or action, or "knowing which." This builds on action-related knowledge, going from knowing how to lessen impact to knowing the relative benefits of different behaviors and actions. This is taught even less than action-related knowledge.

Attitudes toward nature in this model capture specific attitudes and broader values, measured using the Two Major Environmental Values (2-MEV) Scale (Bogner and Wiseman 2006). According to this model, the environmental value of Preservation of Nature includes attitudes of care toward resources, support of environmental causes, and enjoyment and appreciation of nature. Preservation is frequently labeled an *ecocentric* value. By contrast, the environmental value of Utilization of Nature includes attitudes toward altering nature and human dominance over nature. Utilization is frequently labeled an *anthropocentric* value.

These values are somewhat independent (see **Figure 2**). Many people who care deeply about the environment tend to have high Preservation values and low Utilization values, and many who are uncon-

Promoting Grassroots Action on Environmental Issues
Competence Model for Environmental Education

(Roczen, Kaiser, Bogner & Wilson, 2013)

THE UNIVERSITY OF ARIZONA
College of Education

The Institute for Earth Education

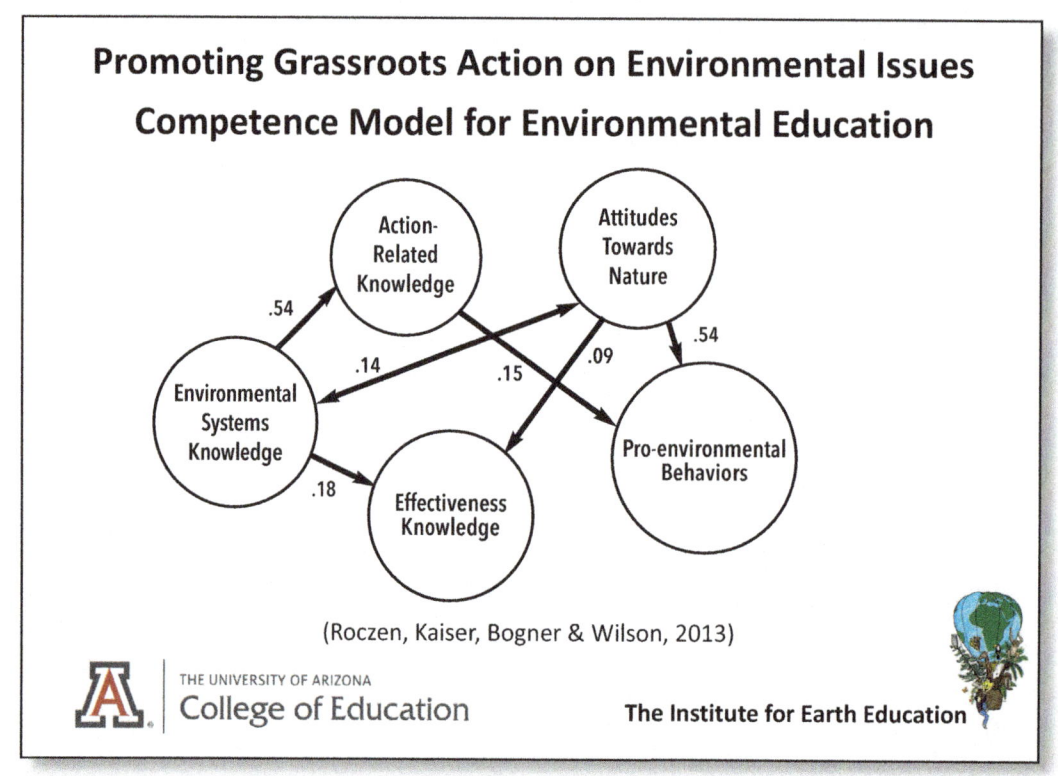

Figure 3

cerned about environmental issues tend to have low Preservation and high Utilization values. However, there are also many people who have both high Preservation and Utilization values; they may believe strongly that the purpose of nature is to support people but believe just as strongly that we must preserve nature. Finally, some people are apathetic toward nature and environmental issues, with low Preservation and Utilization values.

Pro-environmental behaviors in this model are measured by the General Environmental Behavior (GEB) Scale (Kaiser, Oerke, Bogner 2007). Behaviors are categorized as energy conservation, mobility and transportation, waste avoidance, recycling, consumerism, and miscellaneous behaviors toward conservation. It is important to note that the GEB Scale measures self-reported behavior.

Figure 3 shows the results of a statistical test of the model (Roczen et al. 2013). The arrows show relationships between measured constructs in the model; only statistically significant relationships are shown. The number accompanying each relationship arrow shows the strength of the relationship on a spectrum from 0.01 to 0.99. The strongest direct impact on behavior is from attitudes. The parameter, 0.54, indicates a strong relationship. The only other direct predictor of behavior in this model is action-related knowledge, which has a statistically significant but smaller effect

(0.15) on behavior. System knowledge has only indirect influences on behavior, through both action-related knowledge and attitudes. Attitudes also influence system knowledge. People with the strongest self-expressed pro-environmental attitudes tend to have greater system knowledge, along with reporting that they initiate more pro-environmental behaviors. People with greater system knowledge also tend to have greater action-related knowledge, but their pro-environmental behaviors are mediated by their action-related knowledge and their attitudes.

This model clearly demonstrates that knowledge alone is not enough to impact behaviors in a substantive way. Attitudes must be considered. This has important implications for promoting environmental action. People treat knowledge differently based on their attitudes. People who have pro-environmental attitudes are predisposed to pay attention to knowledge about environmental issues. They are more willing to believe, pay attention to, and act on that knowledge. On the other hand, people with neutral or negative environmental attitudes are more likely to be skeptical of information about environmental issues, more likely to ignore it even if they believe it, and less likely to take action. In addition, attitudes influence knowledge; those with pro-environmental attitudes are more likely to seek out information about environmental issues.

Education programs, both formal and informal, can successfully influence environmental attitudes, as will be discussed later in this paper. In addition, messages asking for environmental action can be tailored to achieve more success by paying attention to what research has shown us about values and environmental identity.

Values and environmental identity

People's sense of who they are, their identity, influences the decisions they make. People's values and life goals are one aspect of human identity that plays a significant role in the development of pro-environmental attitudes and behavior (Crompton and Kasser 2009). Values and life goals are critically important to consider when deciding how to convince people to take actions. Psychological research on values has shown that there are universal values, though they play out differently in both individuals and societies (Schwartz 1992). Some goals are extrinsic and are related to self-enhancement; these include achievement, power, status, and wealth (Sheldon and McGregor 2000; Kasser 2005, 2011). Another set of opposing goals are more intrinsic and self-transcendent, including universalism and benevolence (Grouzet et al. 2005; Crompton and Kasser 2009, 2010). People with strong self-enhancement values and goals tend to have negative environmental attitudes and behaviors. Those with strong self-transcendent values and goals tend to be more concerned about the environment and more motivated to engage in pro-environmental behaviors.

Related values tend to reinforce each other. When one is triggered, other, closely related values are also triggered, and opposing values are discouraged. This dynamic has important implications for promoting environmentally positive attitudes and actions. For instance, we might appeal for people to take a particular environmental action, such as using less electricity, for different reasons; we might argue that it will save money or that it will reduce emissions from coal-burning power plants. Both are true, but they trigger different values. Promoting the idea of using less electricity to save money might work quite well, but because it triggers self-enhancing values, it can reinforce values that are associated with negative environmental attitudes and behaviors and so is unlikely to lead to other pro-environmental attitudes and actions. On the other hand, emphasizing how reducing emissions will help the health of those who live near power plants triggers self-transcendent values, which are more likely to lead to greater environmental concern and other pro-environmental attitudes and actions (see Crompton and Kasser 2009, 2010).

What we know from research on universal values clearly has implications for how to successfully promote environmental action. Importantly, the notion of universal values means everyone embraces all these values to some extent. Messages that trigger self-enhancement values are going to do that at least to a small extent even in those with strong environmental values. But emphasizing these messages too heavily can potentially lessen the general pro-environmental actions of those people. Conversely, strongly emphasizing self-transcendent values can potentially lead to more pro-environmental action even in those with low pro-environmental values.

Earth education:
A programmatic approach to environmental learning

Earth education grew out of the groundbreaking work of Steve Van Matre in his Acclimatization summer camp programs to help young people build a love affair with the Earth (Van Matre 1972). An international nonprofit educational organization, The Institute for Earth Education (see www.ieetree.org) was formed in 1974 and continues to this day. The purpose of earth education is the process of "helping people to live more harmoniously and joyously with the Earth and its life" (Van Matre 1990, 87).

Earth education takes a programmatic approach to learning. Earth education programs are magical learning adventures, holistic programs designed to help learners construct understandings of the systems of life that support us, develop positive feelings for the natural world and our place in it, and begin to craft lifestyles that lessen impact. There are several earth education programs for different ages and contexts, including Sunship Earth (Van Matre 1977), Earthkeepers (Van Matre and Johnson 1988), Sunship III (Van Matre and Johnson 1997), and Rangers of the Earth (Van Matre and Farber 2005). Other programs are in development. All are focused on:

- helping participants construct understandings of fundamental ecological concepts:
  o flow of energy
  o cycling of materials

Promoting Grassroots Action on Environmental Issues

Sunship III: Commencement Exercises

Solarians -- units of sunlight energy

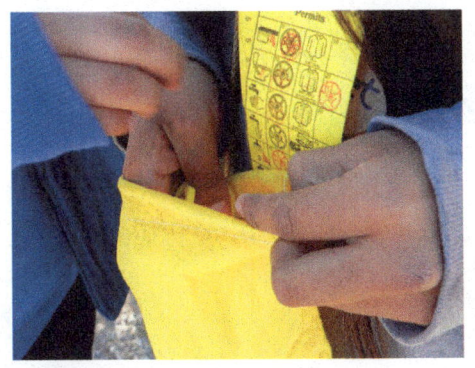

Daily Lifestyle Analysis
- Tracking daily energy use
- Making choices for the next day

THE UNIVERSITY OF ARIZONA
College of Education

The Institute for Earth Education

Figure 4

- o interrelating of life
- o changing of forms
- developing positive feelings in participants such as:
  - o joy at being in touch with the elements of life
  - o kinship with all living things
  - o reverence for natural communities
  - o love for the Earth
- processing what participants learn and experience into action through:
  - o internalizing understandings for how life works on the Earth
  - o enhancing feelings for the Earth and its life
  - o crafting more harmonious lifestyles
  - o participating in environmental planning and action

All earth education programs include substantial time for participants to experience nature firsthand through highly participatory activities.

As an example, Sunship III: Perception and Choice for the Journey Ahead, is designed to help young adolescents begin to craft lifestyles that will have less impact on the natural systems of our

Figure 5

planet. The program is designed around the nature of young people at this stage of life: becoming more independent, making more decisions, and interacting with peers in new ways. The goals of the program are to:

- celebrate the transition from childhood into adulthood
- focus on Perception and Choice in daily habits and routines
- understand how energy and materials tie everything together
- experience good feelings when in nature
- undertake personal lifestyle changes

Participants are invited to a three-day residential "Commencement Exercises," away from school and home, during which they use Solarians (units of sunlight energy) to pay for all the energy and materials they use (see **Figure 4**). They are challenged to change their perceptions of their relationship to the natural world and to make choices that lessen their impact on the systems of life. To learn about the ecological systems that support us, four ecological concepts are explored through outdoor, participatory activities designed to bring these abstract concepts into the realm of the concrete. For instance, to learn about energy, participants visit "Solarville" to order a pizza, discovering the often hidden ways energy is used in our

Promoting Grassroots Action on Environmental Issues

Sunship III: Feelings Activities

Magic Spots

Objet Trouvé

Endangered Species Ceremony

THE UNIVERSITY OF ARIZONA
College of Education

The Institute for Earth Education

Figure 6

daily lives. To learn about the cycling of materials, they become workers at the Cycle Factory, operating the air, water, and soil cycles. Feelings are enhanced through activities like "Objet Trouvé," an exhibit of nature's art that participants create as well as view, and Magic Spots, which is solitude time in touch with the elements of life (**Figures 5 and 6**).

Applying their new understandings, perceptions, and choices in their lives back at home and school is the focus of the "Quest" they embark on after the three-day "Commencement Exercises," seeking truth, adventure, and harmony (see **Figure 7**). Participants work together in small sharing-circle groups to support each other. For instance, they interview role models in their community who are using energy and materials wisely, demonstrating care for natural places and things, and developing a deep personal relationship with the Earth. The goal is to broaden their understandings of options as they craft their own lifestyles.

Research on Earth education programs has investigated changes in participants' understandings, values and attitudes, and proenvironmental behaviors. Results consistently demonstrate increased understanding, strengthened proenvironmental attitudes, and more environmentally positive behaviors. Results also confirm the power of attitudes in the Competence Model for Environmental Education (see Johnson and

Figure 7

Manoli 2008; Baierl, Johnson, and Bogner 2011; Činčera and Johnson 2013; Manoli et al. 2014; Johnson and Činčera 2015; Manoli et al. 2019).

Summary

Change at all levels, from individuals to societies, is needed if we are to shift the course of our current path of environmental destruction. Individuals can alter their own behaviors, engage in group actions, and vote for candidates and policies that will lessen our impacts on the environment. Many people across the globe have taken these and other steps. How do we encourage even more to do so?

Education is a key, but it must be more than just formal schooling. Education can also help us look at how we encourage people of all ages to become more environmentally concerned and involved through informal education, public information campaigns, and local grassroots efforts. To be successful, we need to keep in mind what we know about why people act as they do and how we can encourage proenvironmental views and actions.

First, we must recognize that simply providing information about environmental issues is not enough. We need to pay attention to developing proenvironmental concern and attitudes. We have to help people learn about environmentally positive actions they can take and their effectiveness. Information without the corresponding attitudes and skills is unlikely to result in change.

Second, we must recognize that people's environmental identity—how they see themselves in relation to the environment and nonhuman nature—matters. We can encourage proenvironmental views through appealing to self-transcendent values and life goals, while avoiding reinforcement of self-enhancing values and goals.

Finally, education programs that take a comprehensive approach to addressing ecological understandings, feelings, and behaviors can result in increased knowledge, augmented proenvironmental attitudes, and the adoption of more proenvironmental behaviors. ◊

References

Baierl, T.-M., B. Johnson, and F.X. Bogner. 2021. "Assessing Environmental Attitudes and Cognitive Achievement within 9 Years of Informal Earth Education." *Sustainability* 13 (7): 3622. https://doi.org/10.3390/su13073622

Bogner, F.X., and M. Wiseman. 1999. "Towards Measuring Adolescent Environmental Perception." *European Psychologist* 4 (3): 139–151. https://psycnet.apa.org/doi/10.1027/1016-9040.4.3.139

———, and M. Wiseman. 2006. "Adolescents' Attitudes towards Nature and Environment: Quantifying the 2-MEV Model." *The Environmentalist* 26 (4): 247–254. https://doi.org/10.1007/s10669-006-8660-9

Činčera, J., and B. Johnson. 2013. "Earthkeepers in the Czech Republic: Experience from the Implementation Process." *Envigogika* 8 (4): 1–14. https://doi.org/10.14712/18023061.397

Crompton, T., and T. Kasser. 2009. *Meeting Environmental Challenges: The Role of Human Identity*. Godalming, UK: WWF-UK.

———, and T. Kasser. 2010. "Human Identity: A Missing Link in Environmental Campaigning." *Environment* 52 (4): 23–33. https://doi.org/10.1080/00139157.2010.493114

Grouzet, F.M., T. Kasser, A. Ahuvia, J.M.F. Dols, Y. Kim, S. Lau, R.M. Ryan, S. Saunders, P. Schmuck, and K.M. Sheldon. 2005. "The Structure of Goal Contents Across 15 Cultures." *Journal of Personality and Social Psychology* 89 (5): 800–816. https://psycnet.apa.org/doi/10.1037/0022-3514.89.5.800

Johnson, B., and C. Manoli. 2008. "Using Bogner and Wiseman's Model of Ecological Values to Measure the Impact of an Earth Education Programme on Children's Environmental Perceptions." *Environmental Education Research* 14 (2): 115–127. https://doi.org/10.1080/13504620801951673

———, and J. Činčera. 2015. "Examining the Relationship between Environmental Attitudes and Behaviour in Education Programmes." *Socialni Studia* 12 (3): 97–111. https://doi.org/10.5817/SOC2015-3-97

Kaiser, F.G., B. Oerke, and F.X. Bogner. 2007. "Behavior-Based Environmental Attitude: Development of an Instrument for Adolescents." *Journal of Environmental Psychology* 27 (3): 242–251. https://doi.org/10.1016/j.jenvp.2007.06.004

Kasser, T. 2005. "Frugality, Generosity, and Materialism in Children and Adolescents." *In What Do Children Need to Flourish? Conceptualizing and Measuring Indicators of Positive Development,* edited by K.A. Moore and L.H. Lippman, 357–373. Boston, MA: Springer.

———. 2011. "Cultural Values and the Well-Being of Future Generations: A Cross-National Study." *Journal of Cross-Cultural Psychology* 42 (2): 206–215. https://doi.org/10.1177/0022022110396865

Manoli, C.C., B. Johnson, S. Buxner, and F.X. Bogner. 2019. "Measuring Environmental Perceptions Grounded on Different Theoretical Models: The 2-Major Environmental Values (2-MEV) Model in Comparison with the New Ecological Paradigm (NEP) Scale." *Sustainability* 11 (5): 1286. https://doi.org/10.3390/su11051286

———, B. Johnson, A.C. Hadjichambis, D. Paraskeva-Hadjichambi, Y. Georgiou, and H. Ioannou. 2014. "Evaluating the Impact of the Earthkeepers Earth Education Program on Children's Ecological Understandings, Values and Attitudes, and Behaviour in Cyprus." *Studies in Educational Evaluation* 41: 29–37. https://doi.org/10.1016/j.stueduc.2013.09.008

Roczen, N., F.G. Kaiser, F.X. Bogner, and M. Wilson. 2013. "A Competence Model for Environmental Education." *Environment and Behavior* 46 (8). https://doi.org/10.1177/0013916513492416

Schwartz, S.H. 1992. "Universals in the Content and Structure of Values: Theoretical Advances and Empirical Tests in 20 Countries." *Advances in Experimental Social Psychology* 25: 1–65. https://doi.org/10.1016/S0065-2601(08)60281-6

———. 2006. "Basic Human Values: Theory, Measurement, and Applications." *Revue Française de Sociologie* 47 (4): 249–288. https://doi.org/10.3917/rfs.474.0929

Sheldon, K.M., and H.A. McGregor. 2000. "Extrinsic Value Orientation and 'The Tragedy of the Commons.'" *Journal of Personality* 68 (2): 383–411. https://doi.org/10.1111/1467-6494.00101

Van Matre, S. 1972. *Acclimatization.* Greenville, WV: The Institute for Earth Education.

———. 1977. *Sunship Earth.* Greenville, WV: The Institute for Earth Education.

———. 1990. *Earth Education: A New Beginning.* Greenville, WV: The Institute for Earth Education.

———, and B. Johnson. 1988. *Earthkeepers: Four Keys to Helping Young People Live in Harmony with the Earth.* Greenville, WV: The Institute for Earth Education.

———, and B. Johnson. 1997. *Sunship III: Perception and Choice for the Journey Ahead.* Greenville, WV: The Institute for Earth Education.

———, and L. Farber. 2005. *Rangers of the Earth: Young People Responding to the Planet's Call for Help.* Greenville, WV: The Institute for Earth Education.

Session 3 | First Commentary
on the Presentation by Professor Bruce Johnson

Cultivating Environmental Engagement: Grassroots Collective Action

Professor Dilafruz R. Williams
Leadership for Sustainability Education
Portland State University, Oregon, USA

"Promoting Grassroots Action on Environmental Issues," presented by Dr. Bruce Johnson, is a welcome approach to how we might think about engagement with ecological education issues, given the enormous global environmental concerns impacting all facets of our lives. In his paper, he draws upon long-term research that he and his colleagues have been involved in as they have designed and offered programs on Earth education through The Institute for Earth Education, spanning broad age ranges. In providing one such example with 13–14-year-olds, he makes a compelling case in support of his theories that examine the relationship among knowledge, attitudes, and environmental behavior and also the connection between values and environmental identity.

Among the various points related to environmental issues that Prof. Johnson addresses, I would like to touch upon and expand on the following:

1) First, *providing knowledge/ information alone is not enough.* We need to address attitudes, values, and intentions for any meaningful transformation in behavior. I agree

About the Speaker

- Professor, Leadership for Sustainability Education, Portland State University, USA.
- Researches environmental education, place-based education, garden-based education, and service learning.
- Over three decades of partnerships with communities and schools, including the founding of initiatives such as the Learning Gardens Laboratory and Leadership for Sustainable Education master's program.
- PhD in Cultural Foundations of Education and Curriculum, Syracuse University, USA.

that we must not be fixated on more knowledge related to climate change and environmental problems. Children, youth, and adults are already bodily experiencing the dramatic shifts associated with a rapidly altering climate. Research has shown that knowledge alone is not sufficient for transforming proenvironmental behavior. As Prof. Johnson urges, we need to address values and attitudes to impact changes in behavior.

2) Second, *self-transcendent values rather than self-enhancing values are more effective for encouraging proenvironmental action.* The value of self-transcendence discussed by Prof. Johnson is worth pondering. As an important aspect of understanding how individuals can be motivated to take environmental action, his explanation of intrinsic motivation is right on the mark. Self-awareness of our interconnectedness is of significance especially for action (Zelenski and Desrochers 2021). Sadly, self-transcendent themes are seldom discussed when addressing environmental issues.

3) Third, while education plays a significant role, *formal education is not enough.* Aligned with Prof. Johnson's views, I believe that environmental actions require the expansion of the notion of education beyond the four walls of formal schooling. Informal education, community-based action, and grassroots environmental engagement successfully bring people together to address environmental issues in their own communities. There is power in collective action. Awareness of environmental impacts begins at the local level—in our own backyards, schools, and nearby parks or farms, or being mindful of the quality of the air in our own neighborhoods. Taking care of our own living spaces is a motivator for positive action. The "citizen science" movement, which is the practice of public participation in data collection and research, has parallel objectives and serves diverse populations (Bonney 2021). This is found to have positive impact in raising awareness and successfully engaging the public in advocacy and environmental action.

4) Fourth, I concur with Prof. Johnson's views that programs should support people *to live more harmoniously and joyously with the Earth and its life.*
As David Sobel (2008) reminds us, people protect what they love. In my work on garden education, I have found that cultivating a sense of awe, wonder, and curiosity in children and youth can lead to engagement, kinship, and connection with nature in profound ways (Williams and Brown 2012). My own research on school gardens validates this (Williams 2018). The onslaught of gloom and doom about the

environment and climate issues can cause fear and overwhelm the public and youth, result in despair, and lead to inaction and paralysis. By contrast, love for and attentiveness to one's own place are much more powerful motivators for action. Joyous engagement is more in line with theories of child development in particular but is also applicable to adults.

Changing behavior is a complex process influenced not only by the individual but also by the person's enabling environment. Gender, race, ethnicity, and socioeconomic status also intersect with environmental engagement. Hence, diversity and power inequities would need to be addressed as we explore further how environmental attitudes and behaviors are shaped. Instead of a top-down process, grassroots collective action has played a vital role in the history of social change and increasingly serves to mobilize people of all ages and cultures to address environmental issues.

We can advance this work with further insights from Sander van der Linden (2015) and his colleagues. They have examined best practices for improving public engagement with climate change, drawing upon psychological science, as follows:

- Instead of a distant futurist orientation to climate change, it is more effective to address environmental issues as a "present, local, and personal risk."
- Affective and experiential engagement is more effective, as Dr. Johnson has also argued.
- It is important to leverage social group norms, since group norms affect behavior.
- Gains from immediate action are much more motivating than fear about distant and future losses.
- Tapping intrinsic motivation can support long-term environmental goals.

In conclusion, given the scale of our environmental problems, we need to continue to build on initiatives such as those proposed by Prof. Johnson and ensure that we do so with care and humility. Earth is our true life-giving source. Environmental problems are multifaceted and require a multitude of approaches to address them. Dr. Johnson has enhanced our understanding of the connections among knowledge, attitude, values, and behaviors. By addressing issues *with* others, we can shift our orientation to the powerful collective "we," from the singular "I." ◊

References

Bonney, R. 2021. "Expanding the Impact of Citizen Science." *BioScience* 71 (5): 448–451. https://doi.org/10.1093/biosci/biab041

Sobel, D. 2008. *Childhood and Nature: Design Principles for Educators.* Portsmouth, NH: Stenhouse Publishers.

van der Linden, S., E. Maibach, and A. Leiserowitz. 2015. "Improving Public Engagement with Climate Change: Five 'Best Practice' Insights from Psychological Science." *Perspectives on Psychological Science* 10 (6): 1–6. https://doi.org/10.1177/1745691615598516

Williams, D.R. 2018. "Garden-Based Education." In *Oxford Research Encyclopedia of Education,* edited by G. W. Noblit. New York: Oxford University Press. https://doi.org/10.1093/acrefore/9780190264093.013.188

———, and J.D. Brown. 2012. *Learning Gardens and Sustainability Education: Bringing Life to Schools and Schools to Life.* New York: Routledge.

Zelenski, J.M., and J.E. Desrochers. 2021. "Can Positive and Self-Transcendent Emotions Promote Pro-environmental Behavior?" *Current Opinion in Psychology* 42: 31–35. https://doi.org/10.1016/j.copsyc.2021.02.009

Session 3 | Second Commentary
on the Presentation by Professor Bruce Johnson

Towards Human–Nature Relations that Cultivate Just, Thriving, and Sustainable Worlds

Professor Megan Bang
Learning Sciences and Psychology, Northwestern University, Illinois, USA

Hello, my name is Megan Bang. Today, I would like to speak about learning to cultivate just, sustainable, and culturally thriving communities. This goal is perhaps the greatest challenge of our times, one that will require significant shifts not only in education but also in our values and attitudes toward each other and the Earth. However, today, our formal educational systems across the globe have yet to fully embrace the severity of the challenges in front of us. We have a long way to go in providing the kind of education we need.

As Dr. Johnson's presentation made clear through his work on competence modeling, environmental systems knowledge is necessary but not sufficient. He compellingly argues that people will also need action-related knowledge and knowledge of what is effective. Further, he elevates the need to develop corresponding attitudes and values for meaningful change to occur. He suggests that "We must recognize people's environmental identity, how they see themselves in relation to the environment and nonhuman nature." I could not agree more. Indeed, I applaud Dr. Johnson's work on Earth education that involves increasing

About the Speaker

- Professor, Learning Sciences and Psychology, Northwestern University, USA.
- Mellon Distinguished Scholar, Center for Imagination in the Borderlands, Arizona State University, USA.
- Member, Board on Science Education, National Academy of Sciences, USA.
- Member of Editorial Boards for journals including *Equity & Excellence; Curriculum Inquiry; Journal of the Learning Sciences; Mind, Culture, and Activity; Journal of American Indian Education.*
- PhD in Learning Sciences, Northwestern University, USA.

learners' engagement with pro-environmental activities and attitudes as well as their cultivation of perception and choice in their daily habits.

However, I would like to argue that culture and identity, including the cultural nature of human learning and development, matter significantly for people's environmental identity as well as for the design of effective learning environments. Further, while I appreciate the important goal of fostering young people's search for truth, adventure, and harmony, I would like to suggest that, in order to succeed in such a worthy goal, we also need deeper engagement with the dynamics of power and history and the ways in which these continue to shape people's lives and education.

In my own work with many other colleagues, we have studied what we call *nature–culture relations*. These are fundamental construals between the human world and the natural world. In our work, we argue that these fundamental relations organize human activity. They shape knowledge, systems, and values. They are shaped by power in history. They are foundational to human development and cognition, and they centrally configure learning environments.

Indeed, we have studied two kinds of core cognitive models, what we call the *A Part Of* and the *Apart From* models. A Part Of models see human beings as a part of the natural world, in contrast to Apart From models, which take human beings to be separate or distinct from nature; the latter model embodies a more powered and separatist view or what is sometimes called *human exceptionalism*.

Importantly, Apart From models have come to dominate education in everyday life. Indeed, children's geographies are increasingly indoors, separating them from the natural world. Apart From models shape approaches to content learning; for example, the US is invested in lab-based science infrastructure, not field-based science. We have done studies on how our representational ecosystems are dominated by Apart From models (Medin and Bang 2014): Our books, our media, our diagrams create images of the natural in which human beings are missing. Significantly, there have been a series of studies over the last several decades that demonstrate a substantial decline in the twentieth century in most people's knowledge about plant and animal life (Atran et al. 2004). In short, Apart From models are known to be associated with human-centric reasoning, not whole systems, and unsustainable decision-making.

I see our work as deeply akin to part of Dr. Johnson's studies. However, what I would like to also highlight is that Apart From models are global, but they are differently consequential because of issues of culture, power, and identity. Indeed, most people have no idea that indigenous peoples continue to be in control of a quarter of all Earth's land (Garnett et al. 2018). Indigenous peoples' territories hold 80% of the world's biodiversity, and 95% of the World Wildlife Fund's top 200 climate change hot spots are in indigenous territories.

This matters because in settler colonial nations like the United States, some scholars see that there is a systemic investment in the belittling and shunning—even

"A Part Of" Models of Education
(Aka Land/Water-Based Education, Field-Based Science Education Environmental Education, Outdoor Education)

- Supports socio-ecological systems learning and decision making
- Expands opportunities to engage learners' everyday lives and needs to consequential science learning
- Transforms family and community engagement paradigms
- Driven by "Should we" questions that motivate the need for scientific investigations and ethical deliberation and decision-making
- Time in nature linked to human health (e.g., Frumklin et al. 2017; Tillman et al. 2018)

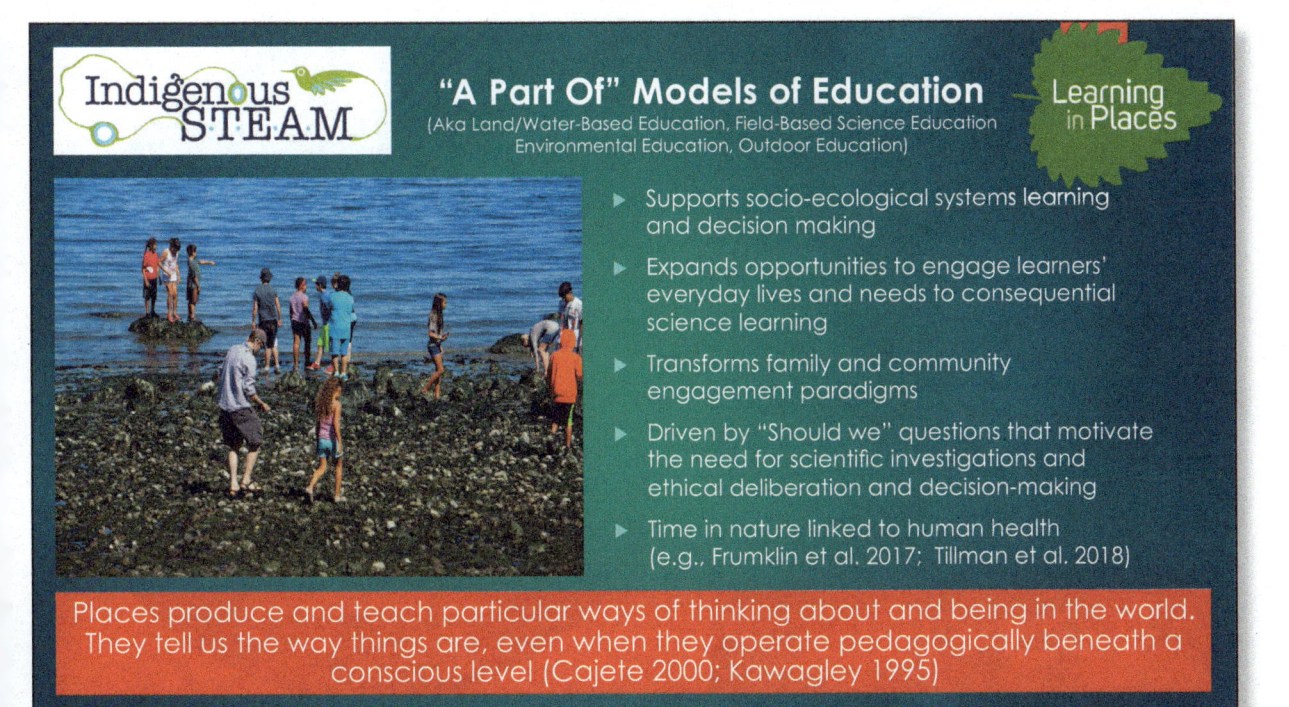

Places produce and teach particular ways of thinking about and being in the world. They tell us the way things are, even when they operate pedagogically beneath a conscious level (Cajete 2000; Kawagley 1995)

Figure 1

erasure—of the value of indigenous peoples' culture. For example, 87% of US educational standards dictate the teaching of indigenous peoples in the context of pre-1900 US history (Shear et al. 2015), if they do it at all. It means that most people have little view of indigenous peoples as related to a beneficial impact on climate change. These kinds of power dynamics persist and are necessary for us to shift.

In our work, we have been developing what we call *A Part Of models of education*. In these models, we are aiming toward supporting socio-ecological systems learning and decision-making. We have developed models that expand opportunities to engage learners' everyday lives and needs to consequential learning. We have worked to transform educational models' relationship to family and communities. We have really worked to develop models of education driven by what we call *Should we* questions that motivate the need for scientific investigations accompanied by ethical deliberation and principled decision-making (**Figure 1**).

Part of this work rests on indigenous models of learning and development that argue that places matter, that where we learn and how we learn matter. In our own efforts, we have been working toward what we call Indigenous STEAM, which is a developed education curriculum that is driven to support the resurgence of indigenous knowledge systems in lifeways for indigenous communities. Another example is a project we call Learning in Places (**Figure 2**).

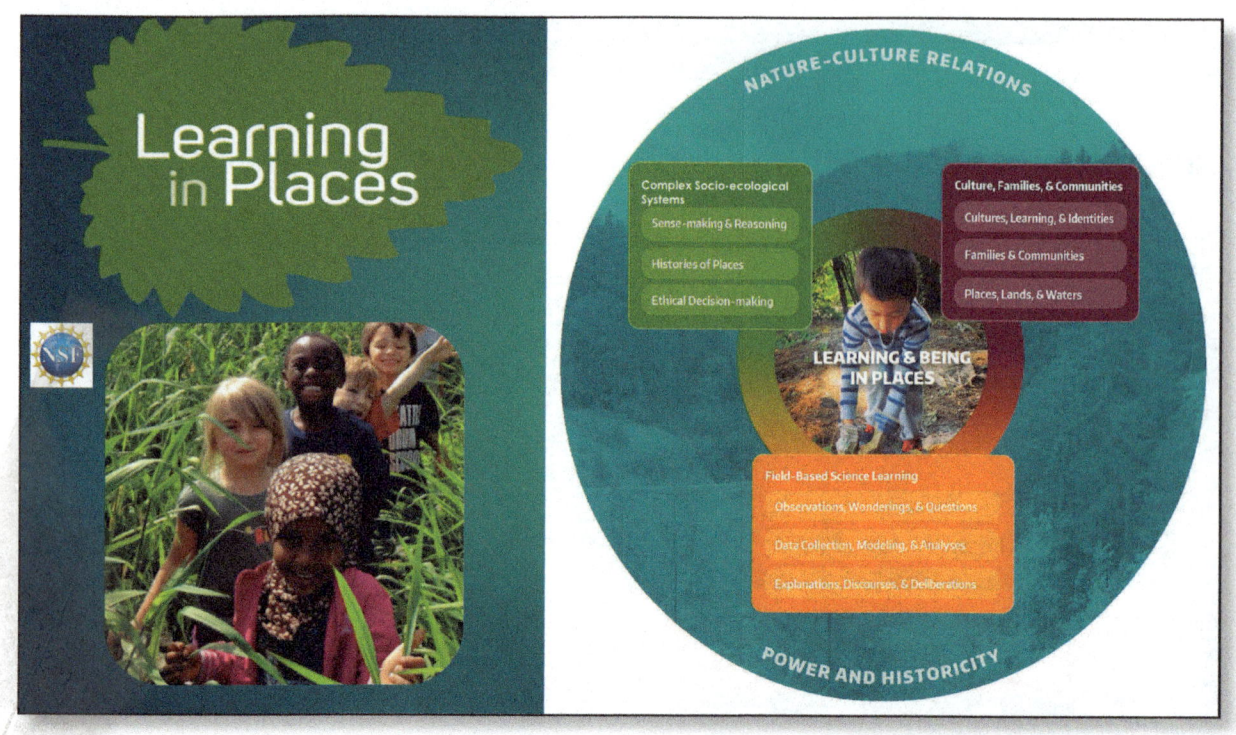

Figure 2

I am going to end my talk here just to suggest that when we develop learning environments, the following elements are always shaping whatever we do with young people: fundamental controls of nature, cultural relations, and power and historicity, as you can see in the diagram on the right in **Figure 2**.

Thus, as we continue toward the important goal of creating educational environments that prepare young people and the next generations in all of our human communities for the challenges that we must learn to adapt to, given climate change, we have to take into consideration culture, power, and historicity.

Thank you very much. ◊

References

Atran, S., D. Medin, and N. Ross. 2004. "Evolution and Devolution of Knowledge: A Tale of Two Biologies." *Journal of the Royal Anthropological Institute* 10 (2): 395–420. https://doi.org/10.1111/j.1467-9655.2004.00195.x.

Cajete, G. 2000. *Native Science: Natural Laws of Interdependence.* Santa Fe, NM: Clear Light Publishers.

Fréchette, B., P.J.H. Richard, P. Grondin, M. Lavoie, and A.C. Larouche. 2018. "Postglacial History of the Vegetation and Climate of the Spruce and Fir Forests of Western Quebec." *Mémoire de Recherche Forestière-Direction de la Recherche Forestière, Ministère des Ressources*

Naturelles, de la Faune 179. Quebec, Canada: Direction de l'Environnement Forestier, Ministère des Ressources Naturelles.

Frumkin, H., G.N. Bratman, S.J. Breslow, B. Cochran, P.H. Kahn Jr., J.J. Lawler, P.S. Levin, P.S. Tandon, U. Varanasi, K.L. Wolf, and S.A. Wood. 2017. "Nature Contact and Human Health: A Research Agenda." *Environmental Health Perspectives* 125 (7). https://doi.org/10.1289/EHP1663

Garnett, S.T., N.D. Burgess, J.E. Fa, Á. Fernández-Llamazares, Z. Molnárm, C.J. Robinson, J.E.M. Watson, K.K. Zander, B. Austin, E.S. Brondizio, N.F. Collier, T. Duncan, E. Ellis, H. Geyle, M.V. Jackson, H. Jonas, P. Malmer, B. McGowan, A. Sivongxay, and I. Leiper. 2018. "A Spatial Overview of the Global Importance of Indigenous Lands for Conservation." *Nature Sustainability* 1: 369–374. https://doi.org/10.1038/s41893-018-0100-6. Accessed on September 14, 2022 https://www.nature.com/articles/s41893-018-0100-6.epdf?author_access_token=ZAToIU uNYxVkZk7d3hQ7M9RgN0jAjWel9jnR3ZoTv0Nlxfg9aDwpfTJNvkjtOhlOfFlXDVJWZFue Kjrvz_ddjYPdyZUDeslOuUlLw0kxM40S57aYeeI-fxx5OnZm1_hkRRK99bTVuwMuTfscdz wTwg%3D%3D

Kawagley, A.O. 1995. *A Yupiaq Worldview: A Pathway to Ecology and Spirit.* Prospect Heights, IL: Waveland Press, Inc.

Medin, D.L., and M. Bang. 2014. "The Cultural Side of Science Communication." *Proceedings of the National Academy of Sciences of the United States of America* 111 (supplement 4): 13,621–13,626. https://doi.org/10.1073/pnas.1317510111

Shear, S.B., R.T. Knowles, G.J. Soden, and A.J. Castro. 2015. "Manifesting Destiny: Re/presentations of Indigenous Peoples in K–12 U.S. History Standards." *Theory & Research in Social Education* 43 (1): 68–101. https://doi.org/10.1080/00933104.2014.999849

Tillman, S., D. Tobin, W. Avison, and J. Gilliland. 2018. "Mental Health Benefits of Interactions with Nature in Children and Teenagers: A Systematic Review." *Epidemiology & Community Health* 72 (10): 958–966. doi:10.1136/jech-2018-210436. Accessed on September 14, 2022. https://jech.bmj.com/content/jech/72/10/958.full.pdf

Session 3
Discussion

Suh-Yong Chung (Chair): I would like to open the floor so that we can take questions or comments from the floor. If you have any questions or comments, please provide them by sending a text note so that the secretary can forward it to me. All right, now, I would like to yield the floor to Prof. Jong Choon Woo.

Jong Choon Woo *(Translation of Korean transcript)*: Yes, hello. Nice to meet you. It was a pleasure to listen to the two presentations and four commentaries today at ICUS. I am Jong Choon Woo, professor emeritus at Kangwon National University. There, I specialized in forestry management for 30 years, lecturing on the subject as well. During this time, I engaged in international projects and traveled to various countries. Forests, which cover 30% of the Earth's total landmass, have a tremendous impact on climate change, and tropical forests in particular are of great importance.

Following the Kyoto Protocol formulated at the end of 1997, various groups determined to resolve the climate change issue, resulting in widespread international discussions. Among the plans that were implemented were the Clean Development Mechanism (CDM) project, Reducing Emissions from Deforestation and Forest Degradation (REDD+), and the Joint Implementation (JI) project. From 2015 to 2017, I went here and there as an evaluator to assess the Korea Forest Service's CDM and REDD+ projects as well as activities in Indonesia and Myanmar.

During my time in Indonesia and Myanmar, I journeyed to those countries' desolated lands, where I helped with restoration and reforestation efforts, and I also aided with the institution of carbon credits. Through interviews with the officials and residents who guided me on my inspections, I came to strongly feel that these projects are absolutely crucial, especially in tropical regions. This was my major takeaway from my one-week assessment survey. Various aspects of the Kyoto Protocol were on full display. At this time, with the Glasgow COP26 conference held last year, despite expecting there to be various changes in policies, I believe that the projects being implemented in actual sites in tropical areas are vitally important.

Among these various projects—the CDM, REDD+, and JI—one thing that is common to them all that is not being properly implemented at this time is the emissions trading system. The emissions trading system is a good system that can be negotiated between developing and developed nations. However, I would like to ask Dr. Matsushita his views on how the emissions trading system can be revitalized. Thank you very much.

Chung: All right, Dr. Matsushita, you may respond to this question.

Kazuo Matsushita: Yes, thank you very much, Prof. Jong Choon Woo, for your very important question. I agree that the afforestation and conservation of forests are very important to attain the goals of the Paris Agreement. As for international cooperation under the Kyoto Protocol, there was a mechanism called the CDM, or Joint Implementation. But under the Paris Agreement, that may be slightly different, because they are aiming at their emissions. All the countries are aiming at their emissions ultimately. We have to be very careful about double-counting the emissions reduction as a result of the afforestation or reforestation.

Of course, it is important to cooperate between tropical forest countries and northern industrial countries to accelerate afforestation and reforestation so that the efforts by these countries are properly credited. However, sometimes there is a sort of overestimation of the amount of CO_2 sequestration as a result of forestation or reforestation. Therefore, we have to be very careful about the calculation of net CO_2 reduction as a result of forest conservation and afforestation. Thank you very much.

Chung: Thank you. If I may add a little bit more comment. In addition to what Dr. Matsushita said, we are talking about implementation of Article 6 of the Paris Agreement. There are similarities but at the same time differences between the systems developed under the Kyoto Protocol, and then systems that will be produced and implemented by the Paris Agreement. In that regard, there are many issues that especially governments and the private sector need to take a look at. Based on what Dr. Matsushita said, especially in the forest sector, countries may need to take a look at the more detailed rules that might be required for the implementation of the systems in their country. Thank you very much. Now, I would like to invite Dr. Takashi Kato for your question. Dr. Kato, you have the floor.

Takashi Kato: I am Takashi Kato from Japan. I would like to speak in Japanese, if I may. I have a question.

(Translation of Japanese transcript) This is a question for Dr. Matsushita. You spoke regarding this concept called *environmental justice*. I have a brief question regarding this. Usually, people think of "environmental justice" as making things convenient for the survival of humans and the human race. I believe that we are supposed to work on the environment because we have reached a point where humanity is suffering damage. At least, it

seems that environmental justice is thought of in that way. I think that in regard to the environment, it is thought of as something that we need to work on once it reaches a state in which it has been damaged by humankind.

One other point is that nature is changing as it is—it is natural and just. There will be cases where it is convenient for humans, and there will be cases where it is not convenient, and nature can change to such a state that nature destroys mankind. But it is Justice for nature. I would be interested to hear what you think about this issue. Thank you very much.

Chung: All right, now. So, the response.

Matsushita: Thank you very much, Kato-san for your very important question, which is rather a theoretical or philosophical question. In terms of environmental justice or climate justice, I think we are using this term to mean a sort of equitable burden sharing. As Prof. Brown said, when an environmental crisis occurs that does not impact everyone worldwide, sometimes it affects poor, vulnerable people more severely than wealthy, privileged people. Thus, we have to pay attention to those who are very vulnerable or who have no capacity to cope with environmental damage. That is, I think, the reason why we are calling for environmental justice.

As for your second point concerning the relationship between human activity and nature, I think it may be different from country to country, from culture to culture. But basically, from the perspective of Japanese citizens, we tend to live in harmony with nature. In order to have a better, healthier life for people, we have to have a healthier and better nature, for nature is the basis of healthy human activity. We have to coexist with nature. That is what I think regarding your question. Thank you very much for your very important question. I think other speakers may have some better responses to your question. Thank you very much.

Chung: All right. Thank you. Maybe we can come back to the issue when we have more time, later in this session. Now, we have another question, from Hiroko Oizumi. The floor is yours.

Hiroko Oizumi: Thank you very much for these excellent presentations and comments. My question is related to Dr. Matsushita. Dr. Matsushita, I missed the initial part of your presentation because of mechanical trouble, so you might have touched upon the issue I am going to ask about now. You said Japan will keep 19% of its coal-fired power? The reason is, I guess, and this is my opinion, nuclear power plants have been shut down because of the accident in 2011. This is only Japan's situation. But all countries have been affected by COVID-19 and the ongoing Ukraine crisis. This is my question: How could COVID-19 and the Ukraine crisis affect and change nations' policies and even goals by 2030? Or by 2050? Whatever it is, I know the fact that in Japan recently, electricity demand was not exceeding supply, and the supply was very safe. Many are concerned about the sanctions on Russian energy resources. That is why I am asking this question. In addition,

I guess all countries should change their energy mix now. Thank you so much.

Chung: All right. Once again, Dr. Matsushita. I find that you are very popular today.

Matsushita: Thank you very much, Oizumi Hiroko-san, for your very important question. I think, as you said, that COVID-19 and the Russian invasion of Ukraine are having a very important impact on the environment as well as human life and human dignity. I think I said in my presentation that we are facing twin problems—COVID-19 and the climate crisis. On top of that, I now have to add military actions. I think war is the greatest form of human and environmental destruction. We have to stop military activities as soon as possible, that is the first thing.

As for the COVID-19 and climate crises, I emphasized the importance of the so-called green recovery. Where many countries are facing a recession from the impact of COVID-19, in order to recover economically, we have to concentrate on investing in climate change–related investments such as renewable energy, energy efficiency, public transportation, and so forth. By investing in such green recovery measures, we will create more green jobs and better working conditions and better lives for people. Also, as for the Russian military invasion in Ukraine, the first thing we have to do is to rapidly transform our energy systems and reduce dependence on fossil fuels like natural gas and coal. We have to expand the use of renewable energy, increase energy efficiency, and develop more regionally based, regional ecologically circular economies. This is a brief answer to your very important question. Thank you very much.

Chung: Thank you. Now, I know we have other wonderful speakers today. Other questions I am sure will be addressed to other speakers as well. In that regard, Prof. Marilyn Brown, the floor is yours.

Marilyn Brown: Thank you very much. I would like to toss this question to Prof. Williams. I very much appreciated the treatment of education, the challenge of transforming attitudes and behavior, beliefs, and the environmental lifestyle transformation we need to encourage. I have followed the research in this area quite diligently, and the theories are very strong. However, I also note from a public policy specialization, that often the education fails to account for the constraints and the barriers that individuals have when they try to realize the lifestyle they would like to have, the environmental assets they would like to purchase, the behavior they would like to pursue. Maybe there is no access to the right infrastructure, maybe you do not have the charging stations, perhaps you do not have the broadband access.

I find in the modeling that there is often a missing element, which is that whole context and how individuals are limited in the realization of what they would like to do. They cannot afford it, there is no program to help buy down the initial heavy capital costs of purchasing a solar rooftop system, you know, all of that. In the education program at the University of Arizona, how do you weave in these social-psychological approaches with some of the realistic constraints that many people have in the different sociodemographic groups that have more constraints than others?

Chung: Okay, professor, the floor is yours for your response.

Dilafruz Williams: I was wondering if your question was directed to Dr. Johnson?

Brown: Yes.

Bruce Johnson: I suspect Prof. Williams could answer similarly to how I would, but I do think it is a really important point. A couple of things I will say about it. One is that we have tested the kind of models of relationships between knowledge, attitudes, values, and behaviors in multiple countries around the world and multiple cultures, and the kind of construct and general relationships hold up quite well. The part, as you know, that has to be very contextualized is the actual environmental behavior part, That is very contextualized. For example, in our research.

The questions we ask depend on the options that are available in that community. It has to be tailored. Secondly, the educational programs absolutely have to be very tailored to the local level, as both Dr. Williams and Dr. Bang discussed. The local level and the place and the community and the culture are extremely important.

For example, in the program I described here, Sunship III for young adolescents, that six-month quest is something the group of learners develops. They have a team of peers, and that sharing circle develops exactly what they will do. To do that, they visit people in their community who are doing things. That is an attempt to try to get them out there to see what is happening in our community, to see how people have found ways of actually doing things, given their situation and limitations and constraints—and to consider what might be some of those ways I might be able to adopt as my own, as I think about my own lifestyle, and others that are simply not going to work.

That does not solve all problems, without a doubt, because there are just structural, societal, and policy issues that really constrain what people can do. I think the only approach we found is to help people craft things for themselves that fit who they are, who their community is, what their culture is, and what the local constraints are. I hope that makes sense. Dr. Williams may have additional thoughts.

Williams: I totally concur with you, Dr. Johnson. As we talked about before, there has to be a multitude of approaches. There is no single standard slate on which you can write solutions, nor can we expect people to act in one particular way that we may think is the only way for us to get results. I think the key is to have communities come together to solve local problems. Plus, we need to empower parents. Dr. Bang talked about this. In some communities, the cultural ramifications of what we do are so absolutely critical. I think that honoring culture is absolutely important, too.

As I mentioned in my response to Dr. Johnson, sustainability is really about getting in connection with *place*. What has happened is that with technology we are more connected to others elsewhere at a distance rather than being present in the moment where we are situated. The local mindset is critical—being engaged with our own "soil" is critical. I call it *living soil*. It is a metaphor. *Where* we are is so important, and to be rooted firmly and take care of our places, is just as important. We need to do so with others, so that we don't feel overwhelmed. It is important to not feel alone in this quest for environmental justice and for sustainability.

Chung: All right. Thank you. We now have one more question coming from the floor, and then, looking at the time that we have, I think, as the chair, I would like to exercise my privilege to ask a follow-up question to Prof. Johnson. I think the education issue is very important, and I am learning a lot from today's presentations and discussions by all of you here. I would like to put that issue in the context of international society, especially with developing countries. My question is, to what extent can we help such countries to actually take the good ideas we are sharing here and incorporate them in their local societies? Prof. Johnson, can you respond to this?

Johnson: That is the trillion-dollar question right there. I have a couple of reactions to that. One is that, as others have noted, we really have not had a serious educational response to the environmental crises we face, we really have not. We pay it lip service. There is a little bit here and there. But really, if you look at the curriculum in almost any country in the world, it takes up a very small part of what we teach and learn about. You know, people in this country often talk about, let us get back to the basics—reading, writing, arithmetic. Well, there is nothing more basic than understanding how our planet's systems work and our impact on them and how we are part of those systems. If we do not have that down, nothing else really matters. But that is a really hard thing to get into education policy.

I think the same is true for developed and developing countries. It is a matter of figuring out how we can get people to really take this seriously. I wish I had an easy answer for that. I do like the fact that there are lots of things people are trying and models that people can take, both in the developed world and the developing world. There is just no easy solution to it. But I think it has to be, again, locally based, in terms of how we get ecology curriculum into the education culture. The other point I want to make, as we have talked about a little bit already today, is that it cannot just be the formal education system; that is always going to be limited.

Therefore, these informal opportunities—the opportunities for people to do things in their communities, with their families, in afterschool programs, informal science education, citizen science, all of those things—turn out to be incredibly powerful, and, in reality, actually have more influence, I think, on what people do than our formal education system.

Chung: Right. Thank you. Now, we have one final question by Dr. Takahiro Hiroi. The floor is yours.

Takahiro Hiroi: Thank you. I think the purpose of this ICUS is, of course, world peace and the happiness of humankind, including that of future children. I wonder if simply reducing CO_2 emissions alone can actually have any effect. As I wrote in my chart, China produces more CO_2 than other developed countries. But because of the lack of free speech and human rights in China, it is very difficult to raise a voice for environmental human rights there. It seems that, however hard free countries like Japan and the US try, there will be only a sort of halfway global effect unless Russia and China and other such countries are liberated from totalitarianism.

Also, even if the global temperature rises 5 or 10 °C, cold regions such as Siberia and Greenland will be producing crop harvests and achieving habitable conditions. I think the most important thing is that because climate change may be inevitable and make some regions inhabitable, we should eliminate national borders so that people can migrate to the better environments by that time. This Earth is really designed to be fed back negatively, so that it is always habitable in various Earth regions. And if there are no boundaries, people will not be prevented from moving around. That is my two cents I wanted to offer.

Chung: I believe that was not a question but rather a personal comment, right? I do not know whether or not it is possible for us to eliminate territorial jurisdictions now, but at least we need to make a broader effort to cooperate with each other so that we can make decisions as if we do not have any territorial boundaries, because climate is just the one single problem common to all of us, and no matter whether we come from the current generation or future generations. I think we do not have any further questions from the floor. Still, we have some time left, so I want to take the liberty of giving one additional minute to each speaker so you can make your final comment or share your final views with all of us. Let me start with Dr. Matsushita. Your final words?

Matsushita: Yes. Thank you very much for all the important questions and comments by the several commentators, and also the honorable chairperson. I would like to share with you what I said in my presentation that the climate crisis is a crisis of human dignity and human survival. However, if we work together, we can achieve a better world and better planet. Therefore, let us work together. Thank you very much.

Chung: All right, his final message is, let us work together. That is very important. Right, thank you. Now I would like to invite Dr. Wil Burns to speak.

Wil Burns: I would want to conclude with a message of some hope. One of the things I would emphasize is that when we embarked 20-plus years ago on international climate regimes, beginning with the Untied Nations Framework Convention on Climate Change, a lot of climate scientists at the time predicted that ultimately global temperatures by the end of the century might rise 4, 5, maybe even

6 °C. When we look at the picture now, if all of the commitments that the parties have made to the Paris Agreement are actually implemented, we are conceivably looking at temperature increases of more like 1.8 °C, which is still certainly serious—right?—as we are already seeing the negative implications of climate change.

But that is just massively different from the posited temperature increases we were looking at before we established these international regimes, and the potentially catastrophic implications for human institutions and natural ecosystems. It provides some hope to me that the international community, through cooperation, can effectuate the kind of changes that minimize negative impacts for future generations. That means we have to keep the pressure on, we have to continue to provide the kind of funding that brings costs down for energy efficiency and decarbonization, so that they become compelling, just from an economic perspective. We can cut through all of the political fog, which we might necessarily need to do. But there is some hope, given the fact that we have substantially bent the curve and showed that we can work together globally to address a global problem.

Chung: Right, thank you. In the end, everything will be about the policy coordination among the different countries. In that regard, I believe that the will of the people, of the state, and of national leaders—presidents and heads of government—is all very, very important. All right, thank you. Now then, I would like to invite the Prof. Marilyn Brown for her final comments.

Brown: Dr. Matsushita, I wanted to compliment you again on your presentation and so much material. I really like the focus on the transformation of innovation into just transition. However, we do not want to leave innovation behind. I felt some concern over the focus on the distant possibilities, such as hydrogen and ammonia, at the expense of what we know to be good short-term innovations and advantages that many economies have taken such advantage of in growing their economic systems, including Japan and the United States.

You know, Japan has been a place for so many advancements that the US has looked at, including in transportation and buildings, and has been a real global leader. But it seems that your current policies are looking to the future at what might be created if we can move to green or blue hydrogen or better ammonia. I know I always am reminding my own government representatives that climate change mitigation and adaptation can be good for the economy, that it can grow, that solutions are good for humanity, and that this is not a competition. This is a win-win approach. Perhaps that is something we can all remind our policymakers of routinely, and hopefully they will absorb that.

Chung: Right. I firmly believe that it is a good solution. It is not a matter of the cost, but a matter of creating new opportunities for all of us. Having said that, I would like to invite Prof. Bruce Johnson for your final comment.

Johnson: I have two very young grandchildren, whom I am really enjoying, but it also puts me in a bit of a dilemma; I am not going to be around after 2050, but they certainly will be, and I worry about what the world is going to be like. However, I am an optimist at heart, and if you want to get some positive energy around any of these issues, go work with some children. They will get you excited; they will renew your hope. They are passionate, they are committed, they are vocal. Talk to university students, talk to youth, talk to really young kids, and they will get you energetic and optimistic about the future.

Chung: Right, thank you. Well, I just found that there is a one commonality that everybody has, because we are all positive about the future. Let us work on it. All right, the final word will be given to Prof. Williams. The floor is yours.

Williams: Wonderful. On a note of hope, I just want to say I have visited hundreds of school gardens. Wherever I go, educators are in favor of outdoor education, getting children and youth connected to nature, getting them outside the four walls of the classroom, breaking asphalt, bringing back soil into mind as it resurfaces from the ground. When soil is visible, we are more likely to be conscious of this life-giving source. I think that is where we start—reconnect with nature where we live—and that is happening.

Plus, youth are demanding action at school board meetings. They are going to school board meetings and asking for climate change curriculum and climate change education. They know what is going on. They are expecting adults to grow up and to address these matters in schools and in their education process. There is hope there.

Chung: All right, thank you. I think this ends our session on engaging the public in tackling environmental concerns. We found that there are common objectives we have, but the only challenge is that we are fragmented. The question is how we can be united, including not only putting policy agendas together but also putting education agendas together, right? That might be also another very important and critical question for us to achieve this goal. Having said that, thank you very much for remaining in the session until now. This session is now closed. Thank you very much. ◇

ICUS XXVIII

CLOSING SESSION

Report on Session 1
Addressing Climate Change: Strategies to Achieve "Net Zero"

Professor Cliff I. Davidson
Session 1 Chair

Thank you very much. I am delighted to speak on the discussions that we had in Session 1, which involved addressing climate change and strategies to achieve net zero. There were two topics within the session. Topic 1 was on tapping hydrogen for electric power and transportation. The main speaker for the session was Prof. David Blekhman from California State University in Los Angeles. He began this session by discussing some of the recent developments with hydrogen-fueled vehicles in Los Angeles, a city which has really been a leader in this area.

In fact, they opened their hydrogen vehicle fueling facility back in 2014. This was the first fueling station in the world to demonstrate accurate metering of the number of kilograms of hydrogen during fueling. The entire state of California currently has 49 hydrogen fueling stations, and actually many more are being built. There are currently more than, I believe, 100 additional stations that are planned throughout the state of California. A lot of progress is being made on hydrogen fueling in California.

Dr. Blekhman also described a student design contest in 2018, which was pretty exciting, because the students developed plans to use wind energy to produce hydrogen for use in cars and trucks and also in trains and boats. Clearly, they were quite creative in extending where hydrogen fueling can be applied.

Dr. Blekhman presented research from Sweden, where the Nilsson Hydrogen House has successfully demonstrated that both heating and electricity can be provided from hydrogen, allowing the house to be positioned entirely off the electrical grid and without the need for gas or oil for home heating. The Nilsson House is basically completely self-sufficient in terms of its energy needs.

There is also work underway in Switzerland that was described, namely using trucks equipped with hydrogen fuel cells, which will have a range of 500 miles between refueling. Hydrogen fuel cells are also being developed to power trains. For example, we saw a locomotive switch engine that is entirely fueled by hydrogen.

There are also many types of small boats that are using hydrogen. In fact, there is currently the development of a full-size cruise ship that will be fueled by hydrogen in its entirety when it is completed.

Many other very exciting changes are taking place in the marine world. There is a unique hydrogen-powered vessel called the *Energy Observer,* which was created back in 2013. It began traveling under its own power in 2014. The crew of the *Energy Observer* first put to sea off the coast of France and traveled along the French coast. Then they planned a project that brought the vessel down to the Mediterranean, and it traveled throughout that body of water. As an additional project, they used the *Energy Observer* to voyage up to the Norwegian Arctic, up to Svalbard and that area, which really demonstrated the tremendous capability of this hydrogen-powered vessel to use only renewable energy to create hydrogen, and then to use that hydrogen to propel itself through the waters.

They are now planning to travel around the world. I believe that they would have already commenced that voyage if not for COVID and some other problems. However, that undertaking is going to take place in the near future. This was all pretty exciting and showed a lot of possibility for hydrogen-fueled vehicles on all terrains.

There were also two commentators who spoke about hydrogen-fuel vehicles. Kazunari Domen, from The University of Tokyo, discussed different ways of producing hydrogen using conventional methods such as fossil fuels, which is what we term *gray hydrogen,* but also using renewable energy sources, termed *green hydrogen.* He mentioned that, really, the challenge we face is an economic one, because it is currently very expensive to produce green hydrogen. In order to be able to have widespread green hydrogen available economically, he stated, we have to have real breakthroughs in the available technology.

The second commentator was Dr. Dave Edlund, who is the chief executive officer of Element 1 Corp. He noted that, for transportation use, we really would like to have a fairly large range of energy—for example, in the megawatt hour range. That is not currently feasible with batteries. Therefore, we need ways to generate hydrogen on board vehicles. He brought up the fact that it is true that fuel cell vehicles are gaining ground, albeit slowly compared with battery-operated electric vehicles, which seem to be gaining popularity very quickly. He also discussed methanol as a precursor for producing hydrogen, which is a far more efficient and safer way than using ammonia for this purpose. This was essentially the presentation on Topic 1.

The next half of the session was devoted to the question of limiting the impact of climate change by developing negative emission technologies to reduce atmospheric carbon (as opposed to just reducing carbon emissions). The main presenter was Dr. Eric Larson from Princeton University. He said there are a number of ways to actively remove carbon from the air. For example, one of the most promising ways is using the principle that plants and plant biomass can capture carbon dioxide and store it. Another is building facilities for bioenergy with carbon capture and storage (BECCS), which he said has a lot of promise. We understand the fundamental physics and chemistry of BECCS. However, at the present time, the methods are not used extensively. There would need to be a lot of construction and a lot of application of new technologies in order to build BECCS facilities around the country.

Another way to reduce atmospheric carbon is direct air capture (DAC) where chemical reactions with CO_2 are used to remove carbon. The carbon can be stored in a number of forms—for example, as carbonate rocks, which was one example given by Dr. Larson. He explained a couple of different DAC methods, such as dry sorbent capture and liquid solvent capture.

Again, we understand the methods, but the actual technologies that would need to be built to use DAC on a large scale generally need to be developed. We are not quite as far along on developing those technologies for DAC as we are for BECCS. Nevertheless, DAC has a lot of promise. Dr. Larson also identified some of the best places around the US as long-term storage locations where the carbon could be taken safely out of the atmosphere and stored. In addition, it can be monitored so that it stays out of the atmosphere. He also showed us five different pathways to achieve net-zero carbon. It was pretty clear that there are major differences in the energy needed to reduce carbon using these different methods and different pathways. Some needed a lot more energy than others. Others, of course, still require technologies to be developed.

We had two commentators for Dr. Larson's presentation. The first was Prof. Steven Chuang from the University of Akron. He noted that storage of carbon dioxide is in fact likely to require significant purification steps to keep it intact—and this purification could be very energy intensive.

Our second commentator was Prof. Larry Baxter from Brigham Young University, who discussed the importance of the initial mole fraction of CO_2 in determining how much energy was needed to separate that CO_2 out for storage. Essentially, he showed a number of graphs that were very convincing that the amount of energy needed for separating out that CO_2, depending on the initial mole fraction.

I see that my time is running out. I will very quickly mention that I think that both of our presenters and all four of the commentators really did a terrific job in addressing climate change strategies to achieve net zero. We learned a lot about hydrogen for electric power and transportation. We learned a lot about negative emission technologies. There is a lot of promise in these methods. There is still, of course, a tremendous amount of work to be done. I think that one theme that came out of all this is that doing this at scale for the planet is going to be a real challenge and requires a lot more work before it can be applied. With that, I really want to thank the two speakers and the four commentators for an outstanding job. ◊

Report on Session 2
Manufacturing Materials for Eco-Friendly Products

Professor Michael K. Stenstrom
Session 2 Chair

We had two excellent presentations from both speakers, and four excellent commentaries on the subjects of their talks, which was, generally speaking, manufacturing eco-friendly products and chemicals. The term *green chemistry* was used frequently. It is a term I have heard many times and which I appreciate.

Hon. Mike Lancaster of the Chemical Industries Association covered a wide range of topics, too many to cite in a short summary, but I can mention a few. He touched on how green chemistry already affects everyday things we do, things that we take for granted. For example, he talked about disinfecting water with chlorine. We think of chlorine as not a friendly chemical. We know of problems associated with it, but we have to remember the great benefit it has provided us.

He notes, and I agree, that water chlorination is one of the major health advances of the twentieth century, citing studies that approximately one-third of the world still does not have access to clean and safe drinking water. It is a topic very familiar to me because it is part of my research, and I have published on it many times. I have Asian students who come to the University of California, Los Angeles (UCLA), which is kind of a gateway from California to the Pacific and Asia, who are interested in learning about technologies to improve the environment and then taking them back to their countries. I think water disinfection ranks right up there with the invention of antibiotics.

Hon. Lancaster gives us the possibility, or even the probability, with examples, that using green chemistry will improve or continue to improve our everyday lives well beyond the longer term, and maybe even in the short term, in ways not immediately recognizable. I am reminded often of the opposing reaction of many California farmers and other farmers and users of technology when they learn of future plans to ban a pesticide or chemical. I can think of examples of hard pesticides, or bioaccumulating pesticides, and volatile compounds in paints and so forth that create air-quality problems, and how these pessimists predicted gloom and trouble because of their loss.

However, we often discover quite the opposite, that good things happen when we require changes from hard chemicals and move more toward green chemistry. Hon. Lancaster's presentation amplifies that and gives us many examples. These changes can provide long-term benefits, but oftentimes immediate, tangible benefits as well.

Prof. James Clark of York University, UK, and Prof. Bimlesh Lochab of Shiv Nadar University in India supported Hon. Lancaster's themes. They amplified the details and cited the need to seek new ways of developing necessary chemicals from sources other than petroleum. Prof. Lochab mentioned how an important chemical in India has petroleum as its main manufacturing source, which will be phased out or reduced in quantity, and how we need to find alternatives—green alternatives—for such chemicals. There are many other examples that we can find in their presentations.

Prof. Michael Shaver of the University of Manchester, UK, taught us about plastics, and I am a member of the plastics generation. I can remember when we had new and more convenient things made of plastics. However, we learned—at least I learned—different things about plastics, namely, that there are some counterintuitive aspects of plastics management that we tend to think of as being progressive. Sometimes, as he points out, recycling plastics is not the best thing to do. Sometimes recycling is not advisable. There are alternatives, one of which is sometimes avoiding plastics. He notes, and I will remember, one particular example of choosing between a plastic cup and a china cup, and how the plastic cup may have disadvantages that the china cup does not have. I think that is an important lesson for us to learn.

We had supporting commentaries by Prof. Mark Miodownik of University College London and Dr. Carly Fletcher of Manchester Metropolitan University, UK. Among the important aspects of successful recycling that all these commentators and speakers supported was involving the public and investigating ways of gaining public support.

We see often, especially here in the United States, how key individuals—at least key individuals in the public arena—take counterproductive or negative attitudes toward environmentally progressive activities, perhaps because our environmental communication is lacking. For example, we in Los Angeles have separate collections for solid waste into three batches: green yard waste, recyclables (plastic and paper), and the rest, which must be landfilled or composted—but it was 20 years in coming.

However, recycling, while it sometimes has its disadvantages, has reduced our need for landfills by about 60%. Therefore, the conclusion I can make from this is that we engineers and scientists have the obligation to communicate not just to our peers, which I think we are good at (and we get rewarded when we communicate with our peers), but also to the public and, most importantly, to our government. I am sure all of you can think about negative reactions from our various government officials to what we think are progressive and better environmental approaches. Most often, members of our government are the most difficult to educate and convince. I am reminded of a book I am now reading on how to talk to a science denier. I believe it would be good reading for all of us.

My conclusion is that a lot of the benefits we can gain, observed by technical folks, scientists, and engineers, require communicating with nontechnical folks. That is my summary. ◊

Report on Session 3
Engaging the Public in Tackling Environmental Concerns

Professor Suh-Yong Chung
Session 3 Chair

Thank you for the introduction. I chaired the third session titled "Engaging the Public in Tackling Environmental Concerns." We just learned that there are important technologies available—hydrogen and then some of the plastic matters and other technologies. Then, even if we might have pretty advanced technology available to be able to address environmental concerns, the question ultimately is how we actually place all of these approaches into the context of policy development and implementation. At the same time, in addition to the active and coordinated role of the government, there is the need to include other stakeholders, including future generations, and to decide what other things to focus on in order to raise their awareness on these important issues, and how we can actually mainstream all these important issues through education.

Therefore, this session focused on the two topics of, first, policy implementation in different countries, and, second, educating on environmental issues. In this regard, we had two presenters, together with four commentators. The presenters included Dr. Kazuo Matsushita and Prof. Bruce Johnson from Japan and the United States, respectively. We also had four commentators—two on the first issue, namely Prof. Wil Burns and Prof. Marilyn Brown, and two on the second topic, Prof. Dilafruz Williams and Prof. Megan Bang.

For our first topic, Dr. Matsushita nicely introduced the current international political landscape on climate change issues. Then he went on to observe how four major players—the European Union, the United States, China, and, of course, Japan, his own country—have developed, incorporated, and implemented climate change themes into their national policies.

We learned that these four economic powerhouses are actually driving to address climate change concerns as much as they can. Of course, there are differences between them. Even in the case of the United States we know that the Biden administration tries to increase the budget to address climate change by passing the Infrastructure Bill, while the European

Union is very eager to implement its own Green New Deal policy, including introducing carbon border adjustment mechanisms and some other important policy measures.

Regarding China, even if we generally have some sort of understanding that they are behind in the climate arena, the country is actually making good efforts to address climate change, according to Dr. Matsushita. However, he indicates that there is still room for China to improve its efforts. Japan, under the new prime minister's leadership is progressing in dealing with climate issues, even if they still have some policy challenges, including how much they can address the important issue of carbon pricing in Japanese society.

Overall, we learned that each country is making its own efforts under its own circumstances, though there were some general policy implications. The bottom line is that we need to make further, constant efforts, through an improved coordination mechanism at the international level, to provide leadership toward producing implementation outcomes in different countries. In that regard, we heard a similar message from the two commentators for Topic 1.

The first issue I would like to share about is the importance of carbon pricing, which is one of the most important policy measures in terms of bringing the technologies into the market. In case of Japan, even if it has carbon pricing measures in place, its carbon tax rate is generally too low and do not yet have a nationwide emissions scheme. The same thing can be found in the United States.

Generally, there was an emphasis on the important role of carbon pricing at not only the international but the domestic level. In this context, there was a specific discussion not only by the commentators and presenters but by observers commenting from the floor about utilizing a carbon market, or Article 6 of the Paris Agreement. We know that there are some differences between the mechanisms that were developed under the Kyoto Protocol and the mechanism to be used under the Paris Agreement. And then we saw that some of the great potential that countries and also private sectors might be able to use with the new mechanisms under the Paris Agreement. Then another important point from our commentators was that there is a need for including all different actors together. In other words, there needs to be a broader participation by related actors.

There is another important element that was emphasized by one of our commentators about dealing with issues of adaptation. In addition, how do we incorporate issues of the ecosystem into our work of realizing a low-carbon society in the end? Overall, we explored the various policy issues in the different countries, and we noted some common implications that can be considered by any other country, not only at the national but at the international level.

As I indicated, Topic 2 focused on issues of education—which even Topic 1 touched on somewhat. The Topic 2 speakers discussed how to ensure the broader participation of different stakeholders through effective education. We can never deny the importance of education.

It is a fact that there has been a lack of, I would say, investment or understanding or real implementation in promoting education to address environmental concerns, including climate change. The Paris Agreement, different from the Kyoto Protocol, has a single, important article about public awareness and education. In that regard, I think that not only the presenters but also the commentators and general participants saw that there will be an increasing importance of the role of education in addressing environmental concerns. The first speaker, Prof. Johnson, especially emphasized that by education we mean not only the formal school system but the many other types of educational systems that may be available in various forms at the local level.

When we talk about education, we have to focus on both of these latter educational aspects. He said that just providing the information will not be sufficient—that there are other aspects that an education system needs to consider. In other words, information without the corresponding attitudes and skills is unlikely to result in changes. Thus, in providing education, we need to consider putting together all the information and attitudes and some of the related skills so that education brings the maximum societal impact to address environmental concerns, including climate change. Then Prof. Johnson introduced some examples of how these efforts can be pursued by universities and local communities.

In the end, he concluded that education programs that address ecological understandings, feelings, and behavior can result in increased knowledge, more pro-environmental attitudes, and adoption of more pro-environmental behaviors. In this sense, there was one additional exchange of views in our question-and-answer session on how we can apply these advanced ideas on environmental education to developing countries and additional developed countries. Of course, it is a billion-dollar question. In any event, we need to continue to share our advanced experiences and systems in the context of increasing the role of education in other parts of the world as well.

In our discussions by the two commentators, we continued to hear a similar understanding on education issues. One of the commentators shared her views that, instead of a distant, futurist orientation to climate change, it is more effective to address environmental issues as present, local, and personal. In other words, while it is always important for us to have a long-term view and long-term strategies, we must make the utmost effort to bring immediate impact to society through education.

All of the session participants agreed with the commentator that experiencing engagement with environmental issues is more educationally effective than simply absorbing knowledge about them. In other words, gains from immediate actions are much more motivating than fear about distant, future losses.

We had a chance to share this among the one-minute final views by the presenters, and we realized that everyone in the session was thinking that addressing environmental concerns through education is not a difficult or negative endeavor. The question is how, in the end, can all societies be orchestrated to put all these great ideas together in a systematic way. I think that is the topic we need to further explore.

During the question-and-answer session, we had several questions from the floor, mainly asking more about the details of the relevant issues. Once again, we concluded our session by hoping that we actually can, in the end, address climate change issues by developing better-coordinated policy mechanisms not only at the international but also at the national level. During the process, education will play a very important role. This is the summary of our session discussion. Thank you very much. ◊

Closing Session
Closing Remarks
Dr. Douglas D.M. Joo
Chair, ICUS XXVIII and HJIFUS

Distinguished scholars, scientists, and guests from all over the world: As we have come to the end of the Twenty-Eighth International Conference on the Unity of the Sciences (ICUS XXVIII), I want to express my profound gratitude to all of you.

This conference could not have come at a better time. We have been seeing warming trends around the globe, plastic pollution and waste, and the failure to educate the public and form effective public policies. We have examined these issues and have explored promising research, new technologies, projects, and approaches to resolve these challenges.

Regarding the achievement of net-zero carbon emissions, we discussed reports on progress being made in two areas: The development of a hydrogen economy using fuel cells for transportation; and the evolution of significant carbon capture industries. In addition, the application of green chemistry with specific catalysts lowers the amount of energy consumed in manufacturing and allows the use of ecofriendly materials to make essential products. To overcome environmental challenges, we also need effective policy and public education.

Around the world, nations are making commitments to achieve net-zero carbon emissions over the next 20–30 years. However, there remain many uncertainties in the achievement of a carbon-neutral status. Green chemistry needs to be accompanied by increased efficiency in manufacturing, the reduction of waste, and the wise use of resources in a circular economy.

We learned that plastics provide many benefits as compared to potential alternatives, but we need to develop effective technologies for their recycling as well as biodegradation and make them as easy as possible for popular use to meet our goals of environmental protection and restoration.

We need to work together. At our end, HJIFUS is committed to serve as a forum in which scientists worldwide can share their research, learn from each other, and formulate public policy recommendations to heal our environment and create a healthy world for future generations.

I wish to thank our keynote speaker, session chairs, presenters, commentators, and general participants for your excellent presentations and discussions, which made this a productive and successful conference. I want to appreciate the iPeaceTV team for their tireless efforts that enabled us to conduct this conference online. I also wish to acknowledge and thank The Washington Times Foundation and *The Segye Times* for their support.

Over the coming months, we will publish the conference proceedings in print and post the presentations on our website. We will distribute print copies to researchers, educational institutions, policy makers, and environmental organizations. Furthermore, we will introduce the contents to a worldwide audience through our online magazine, *The Earth & I.*

In closing, I would like to encourage you to fill out our reflection form and give us your feedback, commenting on ways we could improve and suggesting timely subjects and expert speakers for our future meetings.

I thank you all very much, and may God bless you. ◊

ICUS XXVIII

APPENDIX

Appendix
Organizing Committee

Dr. Douglas D.M. Joo - *Chair*
Chair, HJIFUS

Dr. Dinshaw Dadachanji
Research Director, HJIFUS, USA

Prof. Cliff I. Davidson
Wilmot Professor and Director, Environmental Engineering Program, Syracuse University, USA

Mr. Didier Guignard
Founder and President, Club Science Paris, France

Dr. Chul Hee Han
Professor, Chemistry, Sun Moon University, Korea

Mr. Nobuyuki Iioka
Deputy Director, ICUS Committee, Japan

Dr. Sung Bae Jin
Chair, Hyo Jeong Academic Foundation, Korea

Dr. Frank Kaufmann
Director, Professors World Peace Academy, USA

Dr. Jin Choon Kim
Co-Chair, Cheon Il Guk Academy of Arts & Sciences, Korea

Dr. Theodore Tadaaki Shimmyo
President Emeritus, Unification Theological Seminary, USA

Mr. Glenn Strait
Former Science Editor,
The World & I: Innovative Approaches to Peace, USA

Dr. Frederick Swarts
Life Sciences Editor, New World Encyclopedia, USA

Dr. Jong Choon Woo
Professor Emeritus, Department of Forestry, Kangwon University, Korea

Master of Ceremonies

Dr. Tyler Hendricks
President Emeritus, Unification Theological Seminary, Korea

Staff

Dr. Jun Young Cha, Executive Director, HJIFUS-Korea
Mr. Gregg Jones, Outreach Director, HJIFUS, USA
Ms. Jimin Moon, Assistant Administrator, HJIFUS-Korea
Mr. Shimpuku Uezono, Associate Research Director, HJIFUS, USA
Mr. Ryu-Sung Weinmann, Associate Administrative Director, HJIFUS, USA

* We wish to thank the many additional staff members who helped as volunteers for ICUS XXVIII.

Appendix
List of Attendees

Akiyama, Tomohiro
Lecturer, Kyoto University, Japan
PhD, Environmental Science

Amemiya, Yoshiyuki
Professor Emeritus, The University of Tokyo, Japan
PhD, Engineering

Aoki, Kazuhiro
Professor, Tokyo Medical and Dental University, Japan
DDS

Asakura, Takashi
Professor Emeritus, Tokyo Gakugei University, Japan
PhD, Health Science

Bang, Megan
Commentator, Session 3
Professor, Northwestern University, USA
PhD, Learning Sciences

Baxter, Larry
Commentator, Session 1
Professor, Brigham Young University, USA
PhD, Chemical Engineering

Berg, Staffan
Coordinator, IAAP, USA
DMin, Energy Healing

Blekhman, David
Presenter, Session 1
Professor, California State University Los Angeles, California, USA
PhD, Mechanical Engineering

Brown, Marilyn
Commentator, Session 3
Regents & Brook Byers Professor, Sustainable Systems, Georgia Institute of Technology, USA
PhD, Geography

Bulow, Nancy
Former Executive Assistant,
World Media Association, USA
MA, Religious Education

Burns, Wil
Commentator, Session 3
Visiting Professor, Northwestern University, USA
PhD, International Environmental Law

Cha, Jun Young
Member, Board of Directors, HJIFUS-Korea
Executive Director, HJIFUS-Korea
PhD, Literature

Choi, Jung Chang
Former President, UTI, Korea
PhD, Ministry

Chuang, Steven
Commentator, Session 1
Professor, Polymer Science,
The University of Akron, USA
PhD, Chemical Engineering

Chung, Suh-Yong
Chair, Session 3
Professor, Division of International Studies,
Korea University, Korea
JSD

Clark, James
Commentator, Session 2
Professor, University of York, UK
PhD, Chemistry

Davidson, Cliff
Chair, Session 1
Professor, Civil and Environmental Engineering, Syracuse University, USA
PhD, Environmental Engineering Science

Domen, Kazunari
Commentator, Session 1
University Professor, Office University Professors, The University of Tokyo, Japan
PhD, Chemistry

Edlund, Dave
Commentator, Session 1
CEO, Element 1 Corp, USA
PhD, Chemistry

Fletcher, Carly
Commentator, Session 2
Research Associate, Department of Natural Sciences, Manchester Metropolitan University, UK
PhD, Natural Sciences

Garratt, Dale
Executive Director, UPF New Mexico, USA
PhD, Language, Literacy and Sociocultural Studies

Garratt, Joy
Representative, New Mexico State Legislature, USA
MA, Educational Leadership

Guichard, Jessica
Researcher, University of Plymouth, UK
Advanced Masters, Ocean Renewable Energy

Han, Hoon
Director, New Life Khan Clinic, Korea
MD; PhD, Immunogenetics

Hayashi, Masahisa
Professor Emeritus, Waseda University, Japan
PhD, Public Finance

Hendricks, Tyler
Master of Ceremonies
Theological Consultant,
Family Federation for World Peace and Unification, International Headquarters, Korea
PhD, Religion

Hiroi, Takahiro
Senior Research Scientist, Brown University, USA
PhD, Mineralogy

Ito, Shoichi
Professor Emeritus, Kyushu University, Japan
PhD, Agricultural Economics

Jin, Sung Bae
Member, Board of Directors, HJIFUS-Korea
Chairman, HyoJeong Academic Foundation and Unification Thought Institute International, Korea
PhD, Philosophy of Science

Johnson, Bruce
Presenter, Session 3
Professor, Environmental Learning and Science Education, Georgia Institute of Technology, USA
PhD, Education Psychology

Joo, Douglas DM
Chair, ICUS XXVIII and Organizing Committee
Chair, Board of Directors; Chair, HJIFUS
PhD, Political Science

Joo, Jaewan
Professor, Sun Moon University, Korea
PhD, Theology

Jun, Dong Joon
Pastor, Dongdaemun Family Church, FFWPU, Korea

Kato, Takashi
Professor Emeritus, Graduate School,
Chiba University, Japan
PhD, Theology, Biblical Studies and Comparative Civilizational Studies

Kato, Tetsuji
Graduate Student, The University of Tokyo, Japan

Kim, Gi Eun
Professor, Seokyeong University, Korea
PhD, Chemical and Biological Engineering

Kim, Hi Seong
Graduate Student, Seoul National University, Korea

Kim, Jin Choon
Member, Board of Directors, HJIFUS, Korea
Professor Emeritus,
SunHak Universal Peace Graduate University, Korea
PhD, Physics

Kim, Soo Min
Professor Emeritus, Sun Moon University, Korea
PhD, Political Science

Kinoshita, Takeshi
Professor Emeritus, The University of Tokyo, Japan
PhD, Engineering

Kojima, Koki
Researcher, Global HR Department,
Sunhak UP Graduate University, Korea
MMin

Kuribayashi, Minoru
Associate Professor,
Okayama University, Japan
PhD, Engineering

Lancaster, Mike
Presenter, Session 2
Head of Innovation,
Chemical Industries Association, UK
MPhil, Chemistry

Larson, Eric
Presenter, Session 1
Senior Research Scholar, Andlinger Center for Energy and the Environment, Princeton University, USA
PhD, Mechanical Engineering

Lee, Jae Young
Chair Professor, Seoul University of Buddhism, Korea
PhD, Religious Studies

Lee, Min Seob
Chairman, Diagnomics Inc., Korea
PhD, Biological Sciences

Lochab, Bimlesh
Commentator, Session 2
Professor, Shiv Nadar University, India
PhD, Chemical Sciences

MacMillan, David
Keynote
James S. McDonnell Distinguished University Professor of Chemistry, Princeton University, USA
PhD, Chemistry

Matsushita, Kazuo
Presenter, Session 3
Professor Emeritus, Kyoto University, Japan
MA, Political Economy

Minami, Katsuyuki
Professor Emeritus, Kitasato University, Japan
PhD, Agriculture

Miodownik, Mark
Commentator, Session 2
Professor, Materials and Society,
University College London, UK
PhD, Turbine Jet Engine Alloys

Moon, Yeon Ah
Member, Board of Directors, HJIFUS, USA;
HJIFUS-Korea
Vice Chairman, Universal Peace Federation Korea
PhD, Sociology

Musaraj, Arta
Editor in Chief, Academicus ISJ, Albania
PhD, Management and Economics

Nakamura, Akira
Professor Emeritus, Akita University, Japan
MD; PhD, Medical Science

Nakanishi, Eiichi
Associate Professor, Bukkyo University, Japan
MS, Health Science

Ngatu, Nlandu
Associate Professor, Kagawa University, Japan
MD; PhD, Medical Science

Oh, Hungkuk
Professor Emeritus, Aju University, Korea
PhD, Mechanical Engineering

Ohkado, Masayuki
Professor, Chubu University, Japan
DH

Oizumi, Hiroko
Former Member, House of Representatives, Japan
Master of Administrative Science

Okabe, Satoshi
Professor, Hokkaido University, Japan
PhD, Environmental Engineering

Oku, Akira
Professor Emeritus,
Kyoto Institute of Technology, Japan
PhD, Engineering

Park, Dongsun
Professor, Busan National University, Korea
MSc, International Politics

Park, Koonam
Lab Manager/Research Associate, Dermatology at School of Medicine, Yale University, USA
MSc, Molecular Biology, Immunology

Park, Richard
Director – III, NuclixBio, Inc., Korea
PhD, Biology

Pyun, Young Sik
Professor Emeritus, Convergence Science and Technology, Sun Moon University, Korea
PhD, Industrial Engineering

Rashed, Fatma
Researcher,
Tokyo Medical and Dental University, Japan
PhD, Oral Biology

Rein, Glen
Founder and CEO,
Quantum-Biology Research, USA
PhD, Neurochemistry

Reyes, Hana
Exercise Specialist,
Kodiak Area Native Association, USA

Rigney, Leslie
Executive Director,
UPF USA Las Vegas, USA
BSN, Nurse-Midwifery

Sakamaki, Yoshika
Trainer of Students, UPA,
Sun Hak UP Graduate University, Korea
MA, Pastoral Studies

Sasaki, Tsuyoshi
Professor, Tokyo University of Marine Science and Technology, Japan
PhD, Fisheries Science

Sayers, Jared
President, Azolla Hydrogen, Canada
MBA

Scherban, Vitaly
Enterprise Architect, Avanade, USA
PhD, Technical Science

List of Attendees

Shaver, Michael
Presenter, Session 2
Professor, University of Manchester, UK
PhD, Chemistry

Shimmyo, Theodore
Member, Board of Directors, HJIFUS, USA
President Emeritus,
Unification Theological Seminary, USA
PhD, Theology

Shinoda, Yasunari
Graduate Student,
Tokyo Institute of Technology, Japan

Sisserson, Timothy
UPF Palau Representative, Palau
M Rel Ed

Stenstrom, Michael
Chair, Session 2
Distinguished Professor, Department of Civil
and Environmental Engineering,
University of California, Los Angeles, USA
PhD, Environmental Systems Engineering

Strait, Glenn
Member, Board of Directors, HJIFUS, USA
Former Senior Editor,
The World & I: Innovative Approaches to Peace, USA

Sugimoto, Norihiko
Professor, Keio University, Japan
PhD, Geophysics

Takahashi, Kazutaka
Professor, The University of Tokyo, Japan
PhD, Agriculture

Tanimoto, Shoichi
Researcher, Institute for Molecular Science, Japan
PhD, Theoretical and Computational Molecular
Science

Terasawa, Kunihiko
Associate Professor, Wartburg College, USA
PhD, Philosophy

Valone, Thomas
President, Integrity Research Institute, USA
PhD, Engineering Physics

von Herzen, Brian
Executive Director, Climate Foundation, Australia
PhD, Planetary and Computer Science

Williams, Dilafruz
Commentator, Session 3
Professor, Portland State University, USA
PhD, Cultural Foundations of Education

Wilson, Andrew
Professor, Scripture Studies,
Unification Theological Seminary, USA
PhD, Near Eastern Languages and Civilizations

Woo, Jong Choon
Member, Board of Directors, HJIFUS, USA
Professor Emeritus,
Kangwon National University, Korea
PhD, Forestry Management

Yang, Pyun Seung
Former Professor, Sun Moon University, Korea
PhD, Pastoral Theology

Yao, Shigeru
Professor, Fukuoka University, Japan
PhD, Engineering

Zandvliet, David
Professor, UNESCO Chair, Canada
PhD, Science Education

Appendix
Founders of ICUS and HJIFUS

The Reverend Dr. Sun Myung Moon and the Reverend Dr. Hak Ja Han Moon began life in Holy Marriage in 1960. They based their life on the understanding that the family is the school of love and that all of humanity is one extended family, with God as our common Parent. They wholeheartedly dedicated themselves to practicing their teaching of true love and working to create a peaceful world. Faithful to this divine calling, they came to embody the philosophy of living for the sake of others and their mission as True Parents of humankind. They found that joyfully living for others resembles God's investment in the creation of the world and human beings. Living for the sake of others is the secret to resolving conflict and ending suffering in the world.

Working side by side for more than 50 years, they founded and developed a vast array of organizations devoted to establishing peace, including the Holy Spirit Association for the Unification of World Christianity (HSA-UWC), Family Federation for World Peace and Unification (FFWPU), the Women's Federation for World Peace (WFWP), the Youth Federation for World Peace (YFWP), the Professors World Peace Academy (PWPA), and the Universal Peace Federation (UPF). They founded prominent media organizations such as *The Washington Times* and *The Segye Times,* among many others around the world. In addition, they established international institutions that support education, industry, sports, the arts, and humanitarian projects. Guided by principles that transcend differences between races, religions, and cultures, these organizations have been working to realize a world of interdependence, mutual prosperity, and universally shared values.

During the Cold War Era, the founders emphasized the need to resolve three core arenas of disorder in the world: communism, religious conflict, and the breakdown of the family. Based on their firsthand experiences in war-torn Korea, they saw that the militant atheism and class warfare advocated by communist forces had led to unspeakable atrocities and suffering of millions. In response, the founders built ideological and educational institutions, such as the International Federation for Victory Over Communism (IFVOC), that challenged these forces and offered counterproposals that contributed

ICUS IV, 1975: Rev. Dr. Sun Myung Moon and Rev. Dr. Hak Ja Han Moon welcome participants in New York City.

significantly toward ending the Cold War.

They tackled head-on the dark history of religious conflicts through the International Religious Foundation (IRF), the American Clergy Leadership Conference (ACLC) and UPF's Middle East Peace Initiative (MEPI). By their example and sacrifice, they pioneered an ever-expanding federation of religious leaders working together in all parts of the world to foster a new era of religious harmony among believers from all faiths.

They taught that science is a necessary component in the pursuit of human happiness and progress. In 1972, they founded the International Conference on the Unity of the Sciences (ICUS). Based on their vision of interdisciplinary cooperation, ICUS conferences have brought together scientists and scholars from the natural sciences, social sciences, and the humanities to work collaboratively with a holistic perspective pursuing the unity of the sciences with absolute values—that is, universal values and the fundamental cause and nature of the universe.

After the ascension of Sun Myung Moon in 2012, his wife, Hak Ja Han Moon, took responsibility for these projects with resolve. In honor of her husband's life, she

Rev. Dr. Sun Myung Moon listens to Professor Morton Kaplan, chairman of **ICUS IX** in Miami.

established the Sunhak Peace Prize Foundation to recognize and award righteous heroes of our time who have taken the lead in the aspects of sustainable human development, conflict resolution, and ecological preservation. She introduced many new organizations such as the Hyo Jeong World Peace Foundation (HJWPF) and the World Christian Leadership Conference (WCLC) to expand on previous activities. *Hyo Jeong* means "filial heart"—indicating the teaching of a heart of filial piety toward God. She expanded UPF with initiatives like the International Association of Parliamentarians for Peace (IAPP), the International Summit Council for Peace (ISCP), the Interreligious Association for Peace and Development (IAPD), the International Association of Academicians for Peace (IAAP), the International Media Association for Peace (IMAP), the International Association for Peace and Economic Development (IAED), and the Think Tank project which particularly gathers the wisdom and expertise of leaders in all sectors of society around the world who want to see the peaceful unification of the Korean Peninsula.

ICUS XXIII, 2017: Rev. Dr. Hak Ja Han Moon (*center*) speaks to ICUS XXIII participants about the responsibility of scientists to protect nature. Sitting with her are Dr. Sun Jin Moon (*left*), Chair of the Pacific Rim Education Foundation, and Dr. Yeon Ah Moon (*right*), Chair of Universal Peace Federation, Korea

Among these new organizations, the Hyo Jeong International Foundation for the Unity of the Sciences (HJIFUS) was founded with the mission of solving environmental problems and living in harmony with nature. Rev. Dr. Hak Ja Han Moon has encouraged scientists and academics to pool their knowledge together to effectively address these issues. She has advocated the development of water purification systems for use in Africa and solving fine dust and air pollution in Eastern Asia. She expressed hope that the people of China, using modern technology, might transform the Gobi Desert into a verdant forest. The founders have expressed that to fulfill the mission of environmental protection and restoration, we need to clearly understand that the cosmos is like one body in which we are inseparably connected to the natural world and its fundamental cause. ◊

ICUS III, London, UK, 1974
Left to right: Dr. Kenneth Mellanby, Dr. Nobusige Sawada, Lord Edgar Adrian, and Rev. Dr. Sun Myung Moon.

ICUS XIV, Houston, USA, 1985
In the receiving line, Rev. Dr. Hak Ja Han Moon greets Chairman Kenneth Mellanby.

ICUS XIII, Washington DC, USA, 1984
In the receiving line, Rev. Dr. Hak Ja Han Moon and Mr. Hyo Jin Moon, the founders' son, greet Nobel Laureate Dr. Eugene Wigner.

Appendix
A Brief History of ICUS

The International Conference on the Unity of the Sciences (ICUS) is a series of conferences founded in 1972 by Rev. Dr. Sun Myung Moon and Rev. Dr. Hak Ja Han Moon. The first in this series (ICUS I), held in New York City, had 20 participants from eight countries and diverse disciplines. Chaired by Edward Haskell, it had the theme, "Moral Orientation of the Sciences."

From the beginning, the ICUS founders sought to guide science toward a broader and integrated approach in the search for scientific principles that encourage connections between disciplines, centering on absolute values—that is, values that benefit all of humanity. Though the number of participants in ICUS I was relatively small, the conference produced serious and substantive scholarly output.

On the foundation of this first meeting, great interest in ICUS caused notable growth in both participant numbers and influence. By the time of ICUS II, held a year later in Tokyo, Japan, eminent scientists and scholars from around the world were in attendance, and the number of participants had nearly tripled.

Over the ensuing years, ICUS continued to grow, attracting increasing global interest and expanding the range of disciplines represented, from the natural sciences to the social sciences and the humanities. The meetings were held in 13 major cities of the world, including London, Boston, San Francisco, and Seoul, and themes and topics began to carry over in successive conferences. In this way, areas of importance could be studied more extensively and at greater depth. More than 2,000 scientists, including over 30 Nobel laureates, maintained enduring work around topics such as integrative concepts in the sciences and the value dimensions of scientific research, while exploring such diverse topics as the ethics of gene manipulation, the value of the ocean in sustaining human life, and organization and change in complex systems in nature.

This first era of ICUS, until the year 2000, produced enduring literature in the sciences, including published proceedings, topical books, and hundreds of refereed papers and articles of new research. Despite the diversity of topics covered, the common theme that spanned these conferences was pursuit of the basis for unity of the sciences, steered by absolute values.

By 2000, the founders happily noted that the concept and goals of ICUS had

ICUS XIV, Houston, USA, 1985 *Left to right:* Dr. Kenneth Mellanby, Dr. Alvin Weinberg, and Rev. Dr. Sun Myung Moon have a discussion.

gained a foothold in the academy. Interdisciplinary, collaborative work had become more commonplace, based on the recognition that artificial boundaries need to be transcended in order to pursue the absolute values of seeking the truth and serving the public good. With that, between the years 2000 and 2016, the founders turned their attention to other pressing matters, particularly, international conflicts arising from religious, racial, and cultural differences that led to much suffering. Unfortunately, those years also saw a steep rise in threats to our human future from pollution and destruction of the environment.

In response to these environmental challenges, Rev. Dr. Hak Ja Han Moon reconvened the ICUS series, orienting its mission toward protecting nature and helping regenerate a wholesome planet. Launching the new series, ICUS XXIII was held in Seoul in 2017, with the theme, "Earth's Environmental Crisis and the Role of Science." Since then, an ICUS meeting has been held every year to analyze environmental problems and identify the most effective solutions to them. Topics covered have included climate change, renewable energy technologies, air and water pollution, regenerative farming, sustainable urban infrastructure, circular economy, waste management, and environmental education.

The solutions discussed at these meetings have been based largely on conventional science and technology, which have led to enormous achievements. In recent years, our attention was drawn to examine and experience potentially revolutionary solutions arising from frontier science, opening up a new era of science.

In this manner, ICUS is calling on the great scientists, engineers, and educators of our time to lead the way by which we can repair our planet and create a beautifully healthy environment for future generations. ◊

ICUS XIV, Houston, USA, 1985 *Above:* Reverend and Mrs. Moon enjoy gifts of cowboy hats from the participants.
ICUS XXIII, Seoul, Korea, 2017 *Below:* Participants and Rev. Dr. Hak Ja Han Moon, who gives the Founder's Address.

Appendix
Introduction to HJIFUS

The Hyo Jeong International Foundation for the Unity of the Sciences (HJIFUS) is a nonprofit organization with the mission to protect and restore the Earth's natural environment. It was founded on the inspiration of Rev. Dr. Sun Myung Moon and Rev. Dr. Hak Ja Han Moon, who have elucidated a unique understanding of the relationships between God, humans, and nature.

Our current environmental problems include climate change; pollution of air, land, and water; changes in ecosystems; challenges in generating clean energy; and depletion of natural resources. These issues have detrimental consequences, such as deterioration in the quality of food and adverse effects on the health of humans and other species.

Modern science and technology are powerful tools that shape our world. When used competently and with the right intentions, they benefit both humans and nature. However, when misused through human selfishness, greed, or ignorance, they cause widespread environmental damage.

To help overcome environmental problems stemming from human behavior, HJIFUS promotes character education to improve the human attitude toward nature. On a practical level, the foundation identifies and helps implement well-rounded technological solutions and methods derived from both conventional science and frontier science. In turn, these technologies and how to use them are freely shared among all levels of soci-

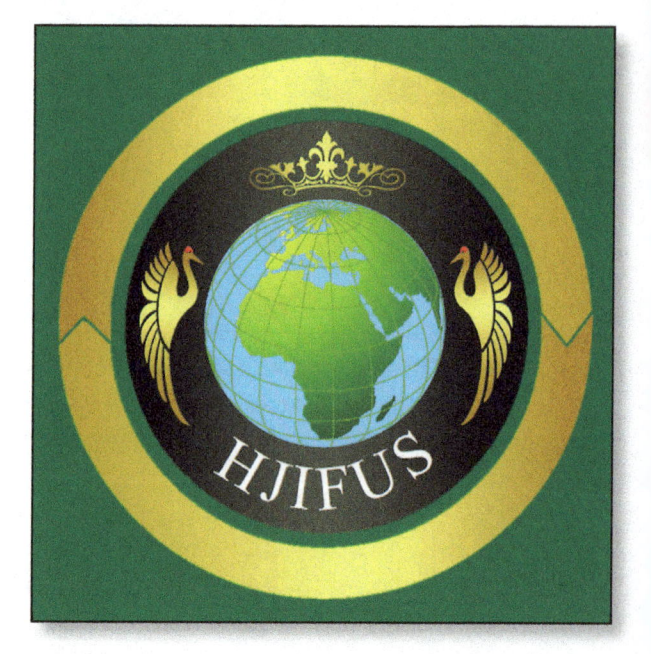

ICUS XXIII participants with HJIFUS founder, Rev. Dr. Hak Ja Han Moon (*front row, center*), February 2017, Seoul, Korea.

ety regardless of race, nationality, religion, social status, or culture. HJIFUS envisions a world where human beings live in transparency as one global family, based on the ideals of interdependence, mutual prosperity, and universally shared values. It is in this way that environmental crises can ultimately be surmounted.

To accomplish these goals, HJIFUS works in six areas. One area of HJIFUS activity is the hosting of international environmental conferences. The International Conference on the Unity of the Sciences (ICUS) was originally started by the founders in 1972 to bring scientists and academics together pursuing the unity of the sciences with absolute values. The ICUS conferences were held regularly until the year 2000. From 2017, HJIFUS reconvened ICUS and has been hosting these conferences annually with a new focus to resolve environmental challenges. These meetings examine technologies derived from conventional science. In addition, HJIFUS has launched another conference series, beginning with the First International Conference on Science and God (ICSG I) in 2020. This series considers environmental solutions offered by frontier science, including postmaterialist science, which may represent a new dawn in the history of science.

In a second area, HJIFUS develops and offers environmental policy proposals to national and international agencies and policy makers. This is accomplished partly by distributing conference proceedings containing such proposals. To

ICUS XXIV, February 2018, Seoul Korea: Rev. Dr. Hak Ja Han Moon (*right*) shares her vision with young scientists, encouraging them to lead future developments in protecting and restoring the Earth environment.

broaden the scope of its activities and to collect consensus from experts and professionals around the world, HJIFUS is considering to establish a World Environmental Council or Congress.

In a third area, HJIFUS produces periodicals and other publications useful for educating the general public on the environment. On Earth Day, April 22, 2021, HJIFUS launched *The Earth & I*, a bimonthly online magazine that provides the latest news on local and global environmental issues, innovative technological solutions, advice on health and life, interviews with leading experts, and human-interest stories and testimonies related to improving the environment. Future publication plans include producing an encyclopedia of environmental science and technology, a *Who's Who* of key people and organizations in the environmental arena, a map of the global environment, and overviews of environmental policies in countries around the world.

The fourth area of HJIFUS activity involves leading social campaigns and programs to promote environmental restoration through service projects, hands-on education, and experiencing nature firsthand so that people of all ages can develop a caring and respectful relationship with the natural world. In July 2021, HJIFUS formed an outreach department to coordinate these activities and to explore partnerships with like-minded organizations.

The fifth area HJIFUS pursues is to raise the living conditions of people through practical projects. From 2018,

HJIFUS representatives visit Senegal and collect water samples for testing, June 2019.

HJIFUS has been laying the groundwork for a project that supplies devices that produce healthy drinking water for communities in need in Africa and elsewhere around the world.

Finally, HJIFUS supports the manufacture of everyday products that can substitute for items that exacerbate environmental problems. Through such products, people can live greener lifestyles and improve their health and well-being while taking care of the Earth.

With these approaches and multi-pronged strategies, HJIFUS is positioned to serve the international community with carefully thought-out plans, innovative technologies, and proven methods to protect, restore, and cherish the Earth for ages to come. HJIFUS organizations have been established in different parts of the world. They operate independently but share the same vision and mission. ◊

Appendix
HJIFUS Boards of Directors

HJIFUS-Korea
(a nonprofit Korean organization)

Dr. Jun Young Cha
Executive Director, HJIFUS-Korea

Dr. Won Ju Chong McDevitt
Chief Secretary of Rev. Dr. Hak Ja Han Moon

Dr. Chul Hee Han
Professor, Chemistry, Sun Moon University

Dr. Sung Bae Jin
Chair,
Hyo Jeong Academic Foundation

Dr. Dong Moon Joo – *Chair*
President, HJIFUS-Korea

Dr. Jin Choon Kim
Co-Chair,
Cheon Il Guk Academy of Arts and Sciences

Dr. Yeon Ah Moon
Chair, Universal Peace Federation, Korea

Dr. Jong Choon Woo
Professor Emeritus, Department of Forestry,
Kangwon National University

HJIFUS (USA)
(a U.S. Section 501(c)(3) organization)

Dr. Michael Jenkins
President, Universal Peace Federation International

Dr. Douglas DM Joo – *Chair*
President, HJIFUS (USA)

Dr. Frank Kaufmann
Director, Professors World Peace Academy, USA

Mr. Larry Moffitt
Executive Vice President,
The Washington Times Foundation

Dr. Yeon Ah Moon
Chair, Universal Peace Federation, Korea

Dr. Theodore Tadaaki Shimmyo
President Emeritus, Unification Theological Seminary

Mr. Glenn Strait
Former Editor, Natural Sciences,
World & I: Innovative Approaches to Peace

Dr. Frederick Allan Swarts
Life Sciences Editor,
New World Encyclopedia

Appendix
Sponsoring and Supporting Organizations and Worldwide HJIFUS Offices

Sponsoring Organizations

Hyo Jeong International Foundation for the Unity of the Sciences-Korea
Hyo Jeong International Foundation for the Unity of the Sciences, USA

Supporting Organizations

The Washington Times Foundation
The Segye Times
Universal Peace Federation International
Family Federation for World Peace and Unification-International

Worldwide HJIFUS Offices

Korea
1st floor, 324-211 Misari-ro, Seorak-myeon
Gapyeong-gun, Gyeonggi-do
Republic of Korea
Tel: +82-(0)70-7771-2604
E-mail: admin.kr@icus.org

United States
3600 New York Avenue, NE
Washington, DC 20002
USA
Tel: +1-202-636-2874
E-mail: admin@hjifus.org

Communication Offices

Japan
2nd Fl., Seiyaku Bldg.
5-13-2, Shinjuku, Shinjuku-ku
Tokyo 160-0022
Japan
Tel: +81-3-5362-0630
E-mail: world@pwpa-j.net

Europe
Espace Barrault
98 rue Barrault
75013 Paris
France
Tel: +33-6-68-31-63-87
E-mail: guignarddidier@gmail.com

Appendix
Index

#

12 Principles of Green Chemistry 93–94
1PointFive 82
2018 Plastics Strategy 136
2021 United Nations Climate Change Conference (COP26) 146, 148–149, 152–153, 155, 158, 163, 186
2-MEV 167

A

absolute values v, 3, 8, 215, 219–220, 223
Academicus International Scientific Journal 140
acrylamide 99
Ahrendt, Kateri 21
Airbus 47
alkylene chain 119
Alstom 45
ammonia 15–16, 57–58, 153, 193, 197
Anastas, Paul 93, 95
anthropocentric 167
Antwerp Port Authority 45
Aquion Energy 105
Aramco 44
Arrhenius equation 82
As a Peace-Loving Global Citizen 11
asymmetric catalysis 18–20, 22
asymmetry 14, 16–17
Audi 34

B

Ballard 45
Bang, Megan 181–182, 190, 201
Barbas, Carlos 22
battery electric vehicle (BEV) 42, 56–58, 85
Baxter, Larry 74, 88–90, 198
Beeson, Teresa 24
Berg, Staffan 85–86
Beyond Benign 95
Biden administration 201
Bio-Plastics Europe 136–137
biobased chemical 107, 112–113
biocatalysis 14, 18–19, 27–28
biodegradable 92, 124, 127–129, 134–137, 142–144
biodiesel 81, 107
bioenergy with carbon capture and storage (BECCS) 30, 63–66, 70–72, 81, 83, 161, 198
biomass 58, 64–65, 68–69, 71, 73, 80–81, 89, 105, 109, 112–114, 197
biorefinery 107–108
Blekhman, David 31, 40–41, 43–44, 50, 52, 55–56, 85–86, 196
blue energy 51
blue hydrogen 51, 193
BMW (Bayerische Motoren Werke) 40
BNSF Railway Company 37, 44
Borths, Chris 22

Brigham Young University 74, 198
Brown, Marilyn 163, 188–190, 193, 201
Build Back Better (BBB) 147
Burns, Wil 157, 192, 201

C

California State University, Los Angeles
    (Cal State LA) 31–44, 46–49, 196
Carbios 101
carbon budget 60–62, 72, 88
carbon capture vi, 30, 51, 64, 74–75,
    77–78, 105, 160, 198, 205
carbon capture and storage (CCS)
    vi, 30, 51, 64, 74–75, 77–78, 105,
    160, 198, 205
carbon capture, utilization, and storage
    (CCUS) 51, 81–82
carbon dioxide removal (CDR) 160–161
carbon farming 63
carbon pricing 153–155, 159, 202
carbon trading 154
carbonate 30, 65, 106–107, 198
carvone 17
cascade catalysis 23, 24
catalysis 4, 13–15, 18–20, 22–24,
    27–28, 50, 94
Cativa process 103
cement 72, 77–78, 106
Chalmers University of Technology 85
Charge Point 36
Chart Industries 74–78, 80
China 111, 151, 154, 158–159, 192,
    200–202, 217
chlorination 97, 199
chlorine 199
Chuang, Steven 81, 198
Chung, Suh-Yong 146, 186–194, 201
circular economy
    4, 6, 113, 129, 135, 205, 220

citizen science 178, 191
Clark, James 111–112, 139, 141, 200
Clean Development Mechanism (CDM)
    186–187
Climeworks 82
coal-fired power generation 148, 153–154
Competence Model for Environmental
    Education 166, 168, 173
Coronavirus disease 2019 (COVID-19)
    3–4, 6, 31, 36, 133, 147, 157–159,
    163, 188–189, 197
cryogenic carbon capture 74
Cummins 35, 46
cyrene (Dihydrolevoglucosenone) 113

D

Daimler 43
Dakar Rally 44
Davidson, Cliff 30, 84–90, 196
decarbonization pillar 70, 81
decarbonize 146–148, 150, 152–157, 159
Declaration on Accelerating the Transition
    to 100% Zero-Emission Passenger
    Cars and Vans 149
Department of Energy (DOE) 48, 52, 54
desalination technology 106, 140
Diels-Alder cycloaddition (DAC) 21–22
direct air capture (DAC) 30, 63, 65, 73–75,
    78–79, 83–84, 89–90, 160, 198
Domen, Kazunari 50, 86, 197

E

Earth education 165–166, 170–171, 173,
    177, 181
ecocentric 167
Edlund, Dave 55, 86, 197
Eldorado 42

electric vehicle (EV) 32, 36, 40, 42, 56, 68, 85–86, 90, 104, 148, 150–151, 197
electrocatalysis 14, 27–28
electrolysis 30, 45, 58, 90, 105
electrolyzer 34–35, 38, 46, 48, 51–52, 58, 105–106
Element 1 55–58, 197
Ellen MacArthur Foundation 136
enantiomer 16–18, 20
Energy Earthshots Initiative 48
Energy Observer 46, 197
environmental justice 151, 187–188, 191
EU Horizon 2020 136
eugenol 121–122
Europe v, 49, 95, 98, 102, 136–137, 159
European Green Deal 149–151
European Union (EU) 49, 58, 86, 97, 100, 136, 149–151, 153–154, 201
Evans, David 18–19

F

Firmenich 25
Fletcher, Carly 135, 142–143, 200
Forbes 31, 40–41, 43–44
France 143, 153–154, 197
fuel cell 31–33, 35–36, 39–40, 42–47, 49, 55, 56–58, 81, 85–86, 90, 150, 196–197, 205
Fuel Cell and Hydrogen Joint Undertaking 49
fuel cell electric vehicle (FCEV) 40, 42, 56–57, 85
Fukushima Hydrogen Energy Research Field 51–52

G

Gaussin 44
General Environmental Behavior Scale (GEB Scale) 168
generic activation mode 21
Georgia 163–164
Giner 35
Glasgow Climate Pact 148, 154
glovebox 19
Government of Japan 153
gray hydrogen 51, 56, 90, 197
green chemistry vi, 4, 92–99, 101–102, 104–105, 108–109, 111, 115, 140–142, 144, 199, 205
Green Deal 136, 149–151
green hydrogen vi, 30, 51–52, 54, 106, 197
Green New Deal 202
green recovery 147, 152, 157–158, 163, 189
greenhouse gas (GHG) 6, 30, 60, 68, 72, 82, 146, 150–152, 155, 158–160
Greenland 192
gross domestic product (GDP) 13, 149, 155
Guichard, Jessica 90

H

Haber, Fritz 15
Harvard University 12, 18
Havila Kystruten 45
Hayashi, Yujiro 22
Hedrick, James 25
High Density Polyethylene (HDPE) 128, 130, 143
Hiroi, Takahiro 87–88, 191–192
Hong, Jeong Kee 6

Hub and Spoke 41
hydrogen economy 31, 37, 48–49, 205
hydrogen energy 50–52
Hydrogen Energy Earthshot 52
hydrogen fuel cell 45, 47, 56, 58, 196
hydrogen fueling station
    35, 38, 41, 56–58, 196
hydrogen hub 31–32, 37, 45, 48–49, 87
Hydrogen Research and Fueling Facility
    31–33, 46
hydrogen train 37, 44
hydrogen valley 31, 49
Hydrogenics 34, 35
Hyo Jeong International Foundation for the
    Unity of the Sciences (HJIFUS)
    v–vi, 3–5, 10, 205, 217–218, 221–225
Hypoint 47
Hyundai Tucson 36

I

imidazolidinone 21–22
India 115, 118–119, 121–122, 144,
    154, 200
Indigenous people 182–183
Institute of Transportation Studies 42
Instron machine 119
Intergovernmental Panel on Climate
    Change (IPCC) 6, 10–11, 60, 72, 89,
    160, 163–164
International Business Machines
    Corporation (IBM) 25
International Conference on the Unity
    of the Sciences (ICUS) v–vi, 2–3, 6,
    8–10, 74, 83, 124, 131, 135, 145, 163,
    186, 192, 205, 214–221, 223–224

International Council of Chemical
    Associations 104
International Energy Agency (IEA) 158
iPeaceTV 4, 206

J

Japan v, 45, 49–52, 86, 147–148, 152–157,
    159, 187–188, 192–193, 201–202,
    215, 219
Johnson, Boris 148, 153
Johnson, Bruce 165, 173–174, 177–179,
    181–182, 190–191, 193, 201, 203
Joo, Douglas D.M. vi, 2, 3, 6, 205
Jørgensen, Karl Anker 22

K

K-Taxonomy 7
Kato, Takashi 187–188
Kawasaki Heavy Industries 45
Kim, Jin Choon 141
kinetic barrier 105
Korea Forest Service 186
Kyoto Protocol 186–187, 202–203

L

Lancaster, Mike 93, 111, 115,
    122, 140–142, 199–200
Lap Shear Strength (LSS) 119
Larson, Eric 60, 74, 80–84, 88–90,
    197–198
lignin 107–108, 115, 119, 123
Linden, Sander van der 179
List, Ben 22
Lochab, Bimlesh 115, 144, 200

Los Angeles (LA) 31–37, 39–49, 56, 92, 139, 144, 196, 199–200

M

MacMillan, David W.C. v, 2, 4, 6, 8, 13
Maersk 59
Manchester Metropolitan University 135, 200
Matsushita, Kazuo 147, 157–160, 163, 187–189, 192–193, 201–202
membrane filtration 98
Merck 25–26
methanol 56–59, 85, 106–107, 197
microplastic 142–143
Ministry of Economy, Trade, and Industry (METI) 52–53
Miodownik, Mark 132, 143, 200
Mirai, Toyota 40
Mitsubishi 45
Molina, Mario 4
Montagnier, Luc 4
Moon, Hak Ja Han v, 2–4, 8–10, 12, 214–224
Moon, Sun Jin 2, 6, 8, 217, 221
Moon, Sun Myung v, 3, 4, 8, 11–12, 214–216, 218–222
Mother of Peace 9
Musaraj, Arta 140–141
Myanmar 186

N

NanoBolt 105
National Institute of Standards and Technology (NIST) 34
National Renewable Energy Laboratory (NREL) 34
natural gas combined cycle (NGCC) 77–78

negative emissions technology (NET) 60, 62–63, 67, 81–83, 160, 197–198
Next Generation EU Recovery Fund 150
Nicewicz, Dave 24
Nilsson house 38, 196
Nilsson, Hans-Olof 38
nitrogen fixation 13, 15–16
Nobel Prize vi, 4, 8, 14, 28
Norway 45, 51, 111
Norwegian Arctic 197

O

Oizumi, Hiroko 86, 188
One Bin 124, 128, 133–135, 137–138, 144
organocatalysis 14, 18, 20, 22, 24–28
Overman, Larry 18

P

Paris Agreement 146, 148, 151–155, 158–159, 187, 193, 202–203
Parking and Transportation Department 36
pesticide 199
petroleum 112, 116–119, 121, 124, 200
phenol 108, 117, 119, 121
photocatalysis 14, 24, 27–28, 50, 52–53, 99, 105–106
photoredox 14, 24, 28
Piper Malibu 47
plastic recycling 101
Plastics Pact Network 136
Plug Power 35
polyacrylamide (PAM) 99
polyethylene (PET) 25, 101, 103, 128, 130
polymer 13, 25, 81, 99, 101–102, 104, 107, 115, 117–119, 121–122, 124, 127–130, 134, 142

PowerCell 45
Princeton University v, 13, 28, 60, 67–68, 197
PEM 32, 35, 45

R

racemic 21
Rangers of the Earth 170, 176
REACH (Registration, Evaluation, Authorisation and Restriction of Chemicals) 95
recycling 92, 101–102, 105, 113–114, 129, 133, 139–140, 167–168, 200, 205
REDD+ (Reducing Emissions from Deforestation and Forest Degradation) 186–187
Regional Clean Hydrogen Hub 48
Regional Hydrogen Clusters 48–49
Request for Information (RFI) 48, 49
reverse osmosis 98
rhodium 159

S

Scania 85
Science Based Targets initiative (SBTi) 97
Shaver, Michael 124, 132–138, 140, 142, 144, 200
Shiv Nadar University 115, 200
Siberia 192
SINTEF (Stiftelsen for industriell og teknisk forskning) 45
Sixth Assessment Report (AR6) 10, 162
Sobel, David 178
social science 12, 140, 215, 219
solventless synthesis 117
sorbent 66, 82–83, 198
space shuttle 32–33, 35

specific energy consumption 102–103
Stanford University 25, 146
Stenstrom, Michael 92, 139–144, 199
Strategic Energy Plan 154
strychnine 24
Suga, Yoshihide 152
Sunship III 170–174, 190
Sustainable Energy Solutions (SES) 74–80
Sustainable Futures 124, 130–131, 138
Sustainable Recovery Plan 158
Svalbard 197
Sweden 31, 38–39, 43, 196
Switzerland 43, 196
synthetic chemistry 14, 142

T

Technology Review 84
telcagepant 26
temperature swing adsorption (TSA) 82–83
Tesla 85
The Earth & I 4, 206, 224
The Glasgow Climate Pact 148, 154
The Global Coal to Clean Power Transition Statement 148
The Segye Times 4, 206, 214, 227
The University of Arizona 165–168, 171–174, 189
The University of Tokyo 50, 197
The Washington Times Foundation 4–5, 11, 206
TMO (2,2,5,5-tetramethyloxolane) 113
Toyota 40, 43, 46, 86

U

U.N. General Assembly 151
Ukraine 86, 159, 188–189

United Kingdom (UK) 93, 111, 115, 124, 129, 132, 135, 139–140, 143–144, 146, 148, 153–154, 157, 200, 218
United Nations Framework Convention on Climate Change (UNFCCC) 146, 148, 154
United States (US) v, 4, 31, 40, 42, 45, 47, 49, 52, 56–58, 60, 66–68, 70–72, 81, 83–85, 87, 93, 151, 154, 158–159, 163, 182–183, 192–193, 198, 200–202
University College London 132, 200
University of Akron 81, 198
University of California, Davis (UC Davis) 42
University of California, Berkeley (UC Berkeley) 20, 157
University of California, Irvine (UCI) 18, 28
University of California, Los Angeles (UCLA) 92, 199
University of Manchester 124, 128, 130, 200
US Pollution Prevention Act 93

V

vacuum-assisted thermal swing adsorption (VTSA) 82–83

Vice-Minister of the Environment 6–7
Vision Motors 43
volatile organic compound 117
Volkswagen 34
Volvo 43, 85

W

Waive 36
Wales 139, 157
Warner, John 93, 95
water chlorination 199
water-splitting 50, 52–54
Williams, Dilafruz 177–178, 189–190, 194, 201
Woo, Jong Choon 186–187
Working Group III (WGIII) 10, 163

XYZ

York University 200
Yorkshire 93, 139
zero-emission vehicle (ZEV) 6, 33, 36, 42
ZeroAvia 47
Zoom 84, 139

Made in the USA
Monee, IL
26 June 2023